Practical Mobile Forensics

Dive into mobile forensics on iOS, Android, Windows, and BlackBerry devices with this action-packed, practical guide

Satish Bommisetty

Rohit Tamma

Heather Mahalik

BIRMINGHAM - MUMBAI

Practical Mobile Forensics

Copyright © 2014 Packt Publishing

All rights reserved. No part of this book may be reproduced, stored in a retrieval system, or transmitted in any form or by any means, without the prior written permission of the publisher, except in the case of brief quotations embedded in critical articles or reviews.

Every effort has been made in the preparation of this book to ensure the accuracy of the information presented. However, the information contained in this book is sold without warranty, either express or implied. Neither the authors, nor Packt Publishing, and its dealers and distributors will be held liable for any damages caused or alleged to be caused directly or indirectly by this book.

Packt Publishing has endeavored to provide trademark information about all of the companies and products mentioned in this book by the appropriate use of capitals. However, Packt Publishing cannot guarantee the accuracy of this information.

First published: July 2014

Production reference: 2140714

Published by Packt Publishing Ltd.
Livery Place
35 Livery Street
Birmingham B3 2PB, UK.

ISBN 978-1-78328-831-1

www.packtpub.com

Cover image by Aniket Sawant (aniket_sawant_photography@hotmail.com)

Credits

Authors
Satish Bommisetty
Rohit Tamma
Heather Mahalik

Reviewers
Dr. Aswami Ariffin
Dr. Salvatore Fiorillo (MSIT)
Yogesh Khatri
Erik Kristensen
Dr. Michael Spreitzenbarth

Commissioning Editor
Rebecca Youé

Acquisition Editor
Rebecca Youé

Content Development Editor
Balaji Naidu

Technical Editor
Manan Badani

Copy Editors
Sarang Chari
Mradula Hegde
Adithi Shetty

Project Coordinator
Aaron.S.Lazar

Proofreaders
Maria Gould
Ameesha Green

Indexer
Hemangini Bari

Graphics
Disha Haria
Abhinash Sahu

Production Coordinator
Adonia Jones

Cover Work
Adonia Jones

About the Authors

Satish Bommisetty is a security analyst working for a Fortune 500 company. His primary areas of interest include iOS forensics, iOS application security, and web application security. He has presented at international conferences, such as ClubHACK and C0C0n. He is also one of the core members of the Hyderabad OWASP chapter. He has identified and disclosed vulnerabilities within the websites of Google, Facebook, Yandex, PayPal, Yahoo!, AT&T, and more, and is listed in their hall of fame.

I would like to thank everyone who encouraged me while producing this book, especially my wife for her great support.

Rohit Tamma is a security analyst working for a Fortune 500 company. His interests lie in mobile forensics, Android application security, and web application security. He is experienced in performing vulnerability assessments and penetration testing of a range of applications, including web and mobile applications. He lives in Hyderabad, India, where he spends time with his parents and friends.

I would like to thank everyone who encouraged me while I was authoring this book, especially my parents and my friends who offered their support in every way they could. Special thanks to Satish Bommisetty, my colleague, co-author of this book, who mentored me all the way through with his valuable suggestions.

Heather Mahalik is the Mobile Exploitation Team Lead at Basis Technology and the Course Lead for the SANS Smartphone Forensics course. With over 11 years' experience in digital forensics, she currently focuses her energy on mobile device investigations, forensic course development and instruction, and research on smartphone forensics.

Prior to joining Basis Technology, Heather worked at Stroz Friedberg and as a contractor for the U.S. Department of State Computer Investigations and Forensics Lab. She earned her Bachelor's degree from West Virginia University. She has authored white papers and forensic course material, and has taught hundreds of courses worldwide for law enforcement, Government, IT, eDiscovery, and other forensic professionals focusing on mobile devices and digital forensics.

There are a lot of people to whom I owe my deepest gratitude. This book is for my husband, who always encourages me to try harder and strive to be one step ahead. This book is also for Jack, who would sleep so that mama could write, and my dad and mother-in-law for always supporting me. Professionally, this book is for Brian Carrier, Eoghan Casey, Terrance Maguire, Rob Lee, and Shawn Howell for getting me addicted to this trade and providing me with the opportunities to better myself. I would also like to thank my co-workers, who have taught me patience, kept a smile on my face, and helped me learn more about forensics than most would deem required. You guys are the best!

About the Reviewers

Dr. Aswami Ariffin specializes in digital forensics (PhD) and previously was a GIAC Certified Forensic Analyst (GCFA) and Certified Wireless Security Professional (CWSP). He has attended various digital forensics training courses, such as SANS System Forensics, Investigation and Response in Australia, multimedia forensics in the United Kingdom and United States, and also data recovery in South Korea.

He has experience in handling computer crimes and computer-related crimes with various law enforcement agencies/regulatory bodies in Malaysia and overseas (recognized as an expert by New South Wales Police Force, Australia). He managed more than 1,800 digital forensic investigations and provided expert testimonies/coordination in Malaysia's High Court and Royal Commission of Inquiry.

He is active in research, and one of his papers entitled *Data Recovery From Proprietary-Formatted Files CCTV Hard Disks* was accepted for publication and presentation at the 2013 Ninth Annual IFIP WG 11.9 International Conference on Digital Forensics, USA. He was also involved as a committee member of the digital forensics program of the prestigious International Conference on Availability, Reliability, and Security (ARES 2012 and 2013).

Due to his immense contribution in combating cyber crimes and developing CyberSecurity, Malaysia's digital forensics capabilities, Dr. Aswami Ariffin was awarded the ISLA (Information Security Leadership Award) in 2009 by ISC2, USA. The Attorney General Chambers of Malaysia and Royal Malaysia Police also issued a commendation letter and certificate of appreciation to him.

Currently, he is Vice President of Cyber Security Responsive Services at CyberSecurity Malaysia. He provides input on strategic direction, technical leadership, and marketing strategy for CyberSecurity Malaysia security operations and research—Digital Forensics Department, MyCERT, and Secure Technology Services.

Dr. Salvatore Fiorillo (MSIT) is a fast learner, problem solver, and open-minded person. He likes unconventional challenges. Holding a degree in Political Science and a Master's degree in IT Security, his interests are wide ranging, from digital forensic and general hacking, to social, anthropological, statistics, and financial studies. He is a network-centric warfare evangelist and gave a speech at De Vere University Arms in Cambridge (UK) during the 2007 conference organized by the Command and Control Research Program (CCRP) within the Office of the Assistant Secretary of US-Defense (NII). He is also the author of *Theory and practice of flash memory mobile forensics*, a 2009 widespread paper on the limits of digital forensic tools (work cited in the 2014 NIST Guidelines on Mobile Device Forensics).

I would like to thank Lucia Tirino and Monica Capasso for their precious help and support throughout. I would also like to thank the people at Packt Publishing; they are all very professional and nice people.

Yogesh Khatri is an assistant professor teaching computer forensics at Champlain College in Burlington, Vermont. Prior to that, he has had a decade of experience working in industry as a consultant and trainer for various companies, including guidance software, during which he worked on cases in several countries, and with many Fortune 100 companies. Yogesh has a Master's degree in Computer Engineering from Syracuse University. He runs a blog at `www.swiftforensics.com`, which showcases his latest research, scripts, ideas, and videos on computer forensics.

Erik Kristensen holds a Bachelor's degree in Computer Science with over 15 years of experience with computer systems that includes computer security, mobile security, and computer forensics. During his time in the United States Air Force, he specialized in computer security and helped pioneer a mobile security program for the BlackBerry, Android, and iPhone devices. He is currently a GIAC Certified Forensics Analyst (GCFA) and is the primary maintainer of the SANS Investigative Forensics Toolkit (SIFT) for computer forensics. He has a broad range of experience and interests. He enjoys problem solving and thinking out of the box. He is currently the lead DevOps engineer for viaForensics, an advanced mobile security and forensics company.

Dr. Michael Spreitzenbarth worked several years as a freelancer in the IT security sector after finishing his diploma thesis with a major in Mobile Phone Forensics. In 2013, he finished his PhD from the University of Erlangen-Nuremberg in the field of Android Forensics and Mobile Malware Analysis. Since this time, he has been working in an internationally operating CERT. His daily work deals with the security of mobile systems, forensic analysis of smartphones and suspicious mobile applications, as well as the investigation of security-related incidents. Alongside this, he is working on the improvement of mobile malware analysis techniques and research in the field of Android and iOS forensics.

www.PacktPub.com

Support files, eBooks, discount offers, and more

You might want to visit www.PacktPub.com for support files and downloads related to your book.

Did you know that Packt offers eBook versions of every book published, with PDF and ePub files available? You can upgrade to the eBook version at www.PacktPub.com and as a print book customer, you are entitled to a discount on the eBook copy. Get in touch with us at service@packtpub.com for more details.

At www.PacktPub.com, you can also read a collection of free technical articles, sign up for a range of free newsletters and receive exclusive discounts and offers on Packt books and eBooks.

http://PacktLib.PacktPub.com

Do you need instant solutions to your IT questions? PacktLib is Packt's online digital book library. Here, you can access, read and search across Packt's entire library of books.

Why subscribe?

- Fully searchable across every book published by Packt
- Copy and paste, print and bookmark content
- On demand and accessible via web browser

Free access for Packt account holders

If you have an account with Packt at www.PacktPub.com, you can use this to access PacktLib today and view nine entirely free books. Simply use your login credentials for immediate access.

Table of Contents

Preface

The exponential growth of mobile devices has revolutionized many aspects of our lives. In what is called the Post-PC era, smartphones are engulfing desktop computers with their enhanced functionality and improved storage capacity. This rapid transformation has led to increased usage of mobile handsets across all sectors.

Despite their small size, smartphones are capable of performing many tasks—sending private messages and confidential e-mails, taking photos and videos, making online purchases, viewing salary slips, completing banking transactions, accessing social networking sites, managing business tasks, and more. Hence, a mobile device is now a huge repository of sensitive data, which could provide a wealth of information about its owner. This has in turn led to the evolution of mobile device forensics, a branch of digital forensics that deals with retrieving data from a mobile device. Today, there is a huge demand for specialized forensic experts, especially given the fact that the data retrieved from a mobile device is admissible in court.

Mobile forensics is all about utilizing scientific methodologies to recover data stored within a mobile phone for legal purposes. Unlike traditional computer forensics, mobile forensics has limitations when obtaining evidence due to rapid changes in the technology and the fast-paced evolution of mobile software. With different operating systems and a wide range of models being released into the market, mobile forensics has expanded over the last 3-4 years. Specialized forensic techniques and skills are required in order to extract data under different conditions.

This book takes you through the challenges involved in mobile forensics and practically explains detailed methods on how to collect evidence from different mobile devices with the iOS, Android, BlackBerry, and Windows mobile operating systems.

The book is organized in a manner that allows you to focus independently on chapters that are specific to your required platform.

What this book covers

Chapter 1, Introduction to Mobile Forensics, introduces you to the concept of mobile forensics, core values, and its limitations. The chapter also provides an overview of practical approaches and best practices involved in performing mobile forensics.

Chapter 2, Understanding the Internals of iOS Devices, provides an overview of the popular Apple iOS devices, including an outline of different models and their hardware. The book explains iOS security features and device security and its impact on the iOS forensics approach. The chapter also gives an overview of the iOS file system and outlines the sensitive files that are useful for forensic examinations.

Chapter 3, Data Acquisition from iOS Devices, covers various types of forensic acquisition methods that can be performed on iOS devices and guides you through preparing your desktop machine for forensic work. The chapter also discusses passcode bypass techniques, the physical extraction of devices, and different ways that the device can be imaged.

Chapter 4, Data Acquisition from iOS Backups, provides a detailed explanation of different types of iOS backups and details what types of files are stored during the backup. The chapter also covers logical acquisition techniques to recover data from backups.

Chapter 5, iOS Data Analysis and Recovery, discusses the type of data that is stored on iOS devices and the general location of this data storage. Common file types used in iOS devices, such as plist and SQLite, are discussed in detail so you understand how data is stored on the device, which will help forensic examiners to efficiently recover data from these files.

Chapter 6, iOS Forensic Tools, provides an overview of the existing open source and commercial iOS forensics tools. These tools differ in the range of mobile phones they support and the amount of data that they can recover. The chapter describes the advantages and limitations of these tools.

Chapter 7, Understanding Android, introduces you to the Android model, file system, and its security features. It provides an explanation of how data is stored in any android device, which will be useful while carrying out forensic investigations.

Chapter 8, Android Forensic Setup and Pre Data Extraction Techniques, guides you through the Android forensic setup and other techniques to follow before extracting any information. Screen lock bypass techniques and gaining root access are also discussed in this chapter.

Chapter 9, Android Data Extraction Techniques, provides an explanation of physical, file system, and logical acquisition techniques to extract information from an Android device.

Chapter 10, Android Data Recovery Techniques, explains the possibilities and limitations for data recovery on Android devices. This chapter also covers the process to reverse engineer Android applications to unearth crucial information.

Chapter 11, Android App Analysis and Overview of Forensic Tools, covers various available open source and commercial tools, which are helpful during forensic examination of Android devices.

Chapter 12, Windows Phone Forensics, provides a basic overview of forensic approaches when dealing with Windows Phone devices.

Chapter 13, BlackBerry Forensics, provides forensic approaches to include acquisition and analysis techniques when dealing with BlackBerry devices. BlackBerry encryption and data protection is also addressed.

What you need for this book

The book provides practical forensic approaches and explains the techniques in a simple manner. The content is organized in a manner that allows even a user with basic computer skills to examine a device and extract the required data. A Macintosh, Windows, or Linux computer will be helpful to successfully perform the methods defined in this book. Wherever possible, methods for all computer platforms are provided.

Who this book is for

This book is intended for forensic examiners with little or basic experience in mobile forensics or open source solutions for mobile forensics. The book will also be useful to computer security professionals, researchers, and anyone seeking a deeper understanding of mobile internals. The book will also come in handy for those who are trying to recover accidentally deleted data (photos, contacts, SMS, and more).

Conventions

In this book, you will find a number of styles of text that distinguish between different kinds of information. Here are some examples of these styles and an explanation of their meaning.

Code words in text, database table names, folder names, filenames, file extensions, pathnames, dummy URLs, user input, and Twitter handles are shown as follows: "To view the raw disk images on the iPhone, connect a jailbroken iPhone to a workstation over SSH and run the `ls -lh rdisk*` command."

Any command-line input or output is written as follows:

```
iPhone4:/dev root# ls -lh rdisk*
crw-r----- 1 root operator 14, 0 Oct 10 04:28 rdisk0
crw-r----- 1 root operator 14, 1 Oct 10 04:28 rdisk0s1
crw-r----- 1 root operator 14, 2 Oct 10 04:28 rdisk0s1s1
crw-r----- 1 root operator 14, 3 Oct 10 04:28 rdisk0s1s2
```

New terms and **important words** are shown in bold. Words that you see on the screen, in menus or dialog boxes for example, appear in the text like this: "iOS provides an option **Erase All Content and Settings** to wipe the data on the iPhone."

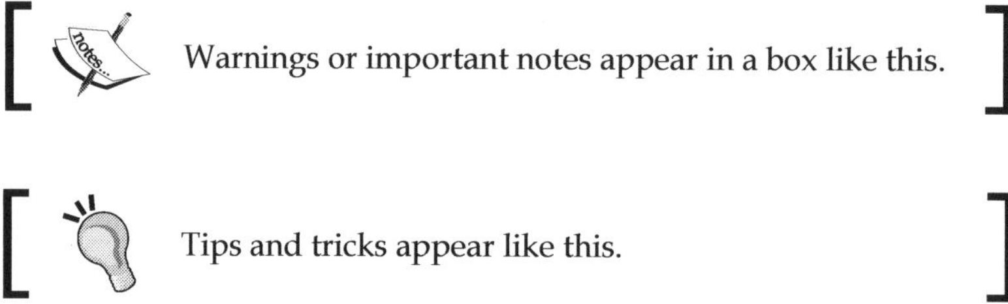

Reader feedback

Feedback from our readers is always welcome. Let us know what you think about this book—what you liked or may have disliked. Reader feedback is important for us to develop titles that you really get the most out of.

To send us general feedback, simply send an e-mail to `feedback@packtpub.com`, and mention the book title through the subject of your message.

If there is a topic that you have expertise in and you are interested in either writing or contributing to a book, see our author guide on `www.packtpub.com/authors`.

Customer support

Now that you are the proud owner of a Packt book, we have a number of things to help you to get the most from your purchase.

Downloading the example code

You can download the example code files for all Packt books you have purchased from your account at `http://www.packtpub.com`. If you purchased this book elsewhere, you can visit `http://www.packtpub.com/support` and register to have the files e-mailed directly to you.

Downloading the color images of the book

We also provide you a PDF file that has color images of the screenshots/diagrams used in this book. The color images will help you better understand the changes in the output. You can download this file from: `https://www.packtpub.com/sites/default/files/downloads/8311OS_ColoredImages.pdf`

Errata

Although we have taken every care to ensure the accuracy of our content, mistakes do happen. If you find a mistake in one of our books—maybe a mistake in the text or the code—we would be grateful if you would report this to us. By doing so, you can save other readers from frustration and help us improve subsequent versions of this book. If you find any errata, please report them by visiting `http://www.packtpub.com/support`, selecting your book, clicking on the **errata submission form** link, and entering the details of your errata. Once your errata are verified, your submission will be accepted and the errata will be uploaded to our website, or added to any list of existing errata, under the Errata section of that title.

Piracy

Piracy of copyright material on the Internet is an ongoing problem across all media. At Packt, we take the protection of our copyright and licenses very seriously. If you come across any illegal copies of our works, in any form, on the Internet, please provide us with the location address or website name immediately so that we can pursue a remedy.

Please contact us at `copyright@packtpub.com` with a link to the suspected pirated material.

We appreciate your help in protecting our authors, and our ability to bring you valuable content.

Questions

You can contact us at `questions@packtpub.com` if you are having a problem with any aspect of the book, and we will do our best to address it.

1

Introduction to Mobile Forensics

In 2013, there were almost as many mobile cellular subscriptions as there were people on earth, says **International Telecommunication Union** (**ITU**). The following figure shows the global mobile cellular subscriptions from 2005 to 2013. Mobile cellular subscriptions are moving at lightning speed and passed a whopping 7 billion early in 2014. Portio Research Ltd. predicts that mobile subscribers will reach 7.5 billion by the end of 2014 and 8.5 billion by the end of 2016.

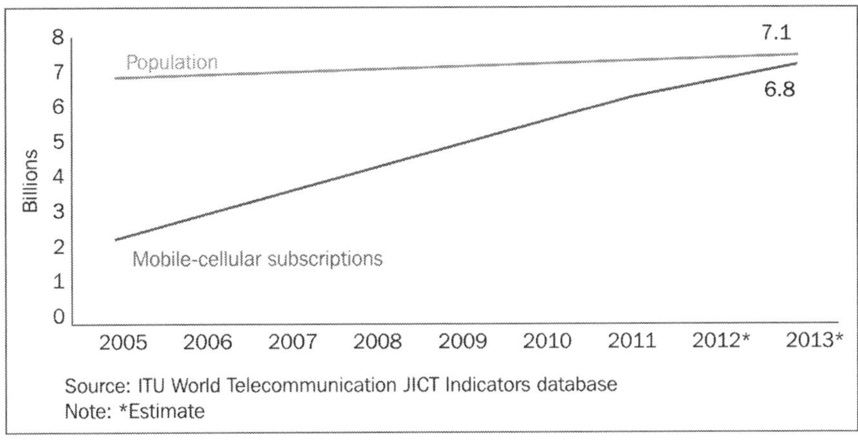

Mobile cellular subscription growth from 2005 to 2013

Smartphones of today, such as the Apple iPhone, Samsung Galaxy series, and BlackBerry phones, are compact forms of computers with high performance, huge storage, and enhanced functionalities. Mobile phones are the most personal electronic device a user accesses. They are used to perform simple communication tasks, such as calling and texting, while still providing support for Internet browsing, e-mail, taking photos and videos, creating and storing documents, identifying locations with GPS services, and managing business tasks. As new features and applications are incorporated into mobile phones, the amount of information stored on the devices is continuously growing. Mobiles phones become portable data carriers, and they keep track of all your moves. With the increasing prevalence of mobile phones in peoples' daily lives and in crime, data acquired from phones become an invaluable source of evidence for investigations relating to criminal, civil, and even high-profile cases. It is rare to conduct a digital forensic investigation that does not include a phone. Mobile device call logs and GPS data were used to help solve the attempted bombing in Times Square, New York, in 2010. The details of the case can be found at `http://www.forensicon.com/forensics-blotter/cell-phone-email-forensics-investigation-cracks-nyc-times-square-car-bombing-case/`. The science behind recovering digital evidence from mobile phones is called **mobile forensics**. Digital evidence is defined as information and data that is stored on, received, or transmitted by an electronic device that is used for investigations. Digital evidence encompasses any and all digital data that can be used as evidence in a case.

Mobile forensics

Digital forensics is a branch of forensic science focusing on the recovery and investigation of raw data residing in electronic or digital devices. Mobile forensics is a branch of digital forensics related to the recovery of digital evidence from mobile devices. **Forensically sound** is a term used extensively in the digital forensics community to qualify and justify the use of particular forensic technology or methodology. The main principle for a sound forensic examination of digital evidence is that the original evidence must not be modified. This is extremely difficult with mobile devices. Some forensic tools require a communication vector with the mobile device, thus standard `write` protection will not work during forensic acquisition. Other forensic acquisition methods may involve removing a chip or installing a bootloader on the mobile device prior to extracting data for forensic examination. In cases where the examination or data acquisition is not possible without changing the configuration of the device, the procedure and the changes must be tested, validated, and documented. Following proper methodology and guidelines is crucial in examining mobile devices as it yields the most valuable data. As with any evidence gathering, not following the proper procedure during the examination can result in loss or damage of evidence or render it inadmissible in court.

The mobile forensics process is broken into three main categories: **seizure**, **acquisition**, and **examination/analysis**. Forensic examiners face some challenges while seizing the mobile device as a source of evidence. At the crime scene, if the mobile device is found switched off, the examiner should place the device in a **faraday bag** to prevent changes should the device automatically power on. Faraday bags are specifically designed to isolate the phone from the network. If the phone is found switched on, switching it off has a lot of concerns attached to it. If the phone is locked by a PIN or password or encrypted, the examiner will be required to bypass the lock or determine the PIN to access the device. Mobile phones are networked devices and can send and receive data through different sources, such as telecommunication systems, Wi-Fi access points, and Bluetooth. So if the phone is in a running state, a criminal can securely erase the data stored on the phone by executing a remote wipe command. When a phone is switched on, it should be placed in a faraday bag. If possible, prior to placing the mobile device in the faraday bag, disconnect it from the network to protect the evidence by enabling the flight mode and disabling all network connections (Wi-Fi, GPS, Hotspots, and so on). This will also preserve the battery, which will drain while in a faraday bag and protect against leaks in the faraday bag. Once the mobile device is seized properly, the examiner may need several forensic tools to acquire and analyze the data stored on the phone.

Mobile device forensic acquisition can be performed using multiple methods, which are defined later. Each of these methods affects the amount of analysis required, which will be discussed in greater detail in the upcoming chapters. Should one method fail, another must be attempted. Multiple attempts and tools may be necessary in order to acquire the most data from the mobile device.

Mobile phones are dynamic systems that present a lot of challenges to the examiner in extracting and analyzing digital evidence. The rapid increase in the number of different kinds of mobile phones from different manufacturers makes it difficult to develop a single process or tool to examine all types of devices. Mobile phones are continuously evolving as existing technologies progress and new technologies are introduced. Furthermore, each mobile is designed with a variety of embedded operating systems. Hence, special knowledge and skills are required from forensic experts to acquire and analyze the devices.

Mobile forensic challenges

One of the biggest forensic challenges when it comes to the mobile platform is the fact that data can be accessed, stored, and synchronized across multiple devices. As the data is volatile and can be quickly transformed or deleted remotely, more effort is required for the preservation of this data. Mobile forensics is different from computer forensics and presents unique challenges to forensic examiners.

Law enforcement and forensic examiners often struggle to obtain digital evidence from mobile devices. The following are some of the reasons:

- **Hardware differences**: The market is flooded with different models of mobile phones from different manufacturers. Forensic examiners may come across different types of mobile models, which differ in size, hardware, features, and operating system. Also, with a short product development cycle, new models emerge very frequently. As the mobile landscape is changing each passing day, it is critical for the examiner to adapt to all the challenges and remain updated on mobile device forensic techniques.
- **Mobile operating systems**: Unlike personal computers where Windows has dominated the market for years, mobile devices widely use more operating systems, including Apple's iOS, Google's Android, RIM's BlackBerry OS, Microsoft's Windows Mobile, HP's webOS, Nokia's Symbian OS, and many others.
- **Mobile platform security features**: Modern mobile platforms contain built-in security features to protect user data and privacy. These features act as a hurdle during the forensic acquisition and examination. For example, modern mobile devices come with default encryption mechanisms from the hardware layer to the software layer. The examiner might need to break through these encryption mechanisms to extract data from the devices.
- **Lack of resources**: As mentioned earlier, with the growing number of mobile phones, the tools required by a forensic examiner would also increase. Forensic acquisition accessories, such as USB cables, batteries, and chargers for different mobile phones, have to be maintained in order to acquire those devices.
- **Generic state of the device**: Even if a device appears to be in an off state, background processes may still run. For example, in most mobiles, the alarm clock still works even when the phone is switched off. A sudden transition from one state to another may result in the loss or modification of data.
- **Anti-forensic techniques**: Anti-forensic techniques, such as data hiding, data obfuscation, data forgery, and secure wiping, make investigations on digital media more difficult.
- **Dynamic nature of evidence**: Digital evidence may be easily altered either intentionally or unintentionally. For example, browsing an application on the phone might alter the data stored by that application on the device.
- **Accidental reset**: Mobile phones provide features to reset everything. Resetting the device accidentally while examining may result in the loss of data.

- **Device alteration**: The possible ways to alter devices may range from moving application data, renaming files, and modifying the manufacturer's operating system. In this case, the expertise of the suspect should be taken into account.
- **Passcode recovery**: If the device is protected with a passcode, the forensic examiner needs to gain access to the device without damaging the data on the device.
- **Communication shielding**: Mobile devices communicate over cellular networks, Wi-Fi networks, Bluetooth, and Infrared. As device communication might alter the device data, the possibility of further communication should be eliminated after seizing the device.
- **Lack of availability of tools**: There is a wide range of mobile devices. A single tool may not support all the devices or perform all the necessary functions, so a combination of tools needs to be used. Choosing the right tool for a particular phone might be difficult.
- **Malicious programs**: The device might contain malicious software or malware, such as a virus or a Trojan. Such malicious programs may attempt to spread over other devices over either a wired interface or a wireless one.
- **Legal issues**: Mobile devices might be involved in crimes, which can cross geographical boundaries. In order to tackle these multijurisdictional issues, the forensic examiner should be aware of the nature of the crime and the regional laws.

Mobile phone evidence extraction process

Evidence extraction and forensic examination of each mobile device may differ. However, following a consistent examination process will assist the forensic examiner to ensure that the evidence extracted from each phone is well documented and that the results are repeatable and defendable. There is no well-established standard process for mobile forensics. However, the following figure provides an overview of process considerations for extraction of evidence from mobile devices. All methods used when extracting data from mobile devices should be tested, validated, and well documented.

A great resource for handling and processing mobile devices can be found at `http://digital-forensics.sans.org/media/mobile-device-forensic-process-v3.pdf`.

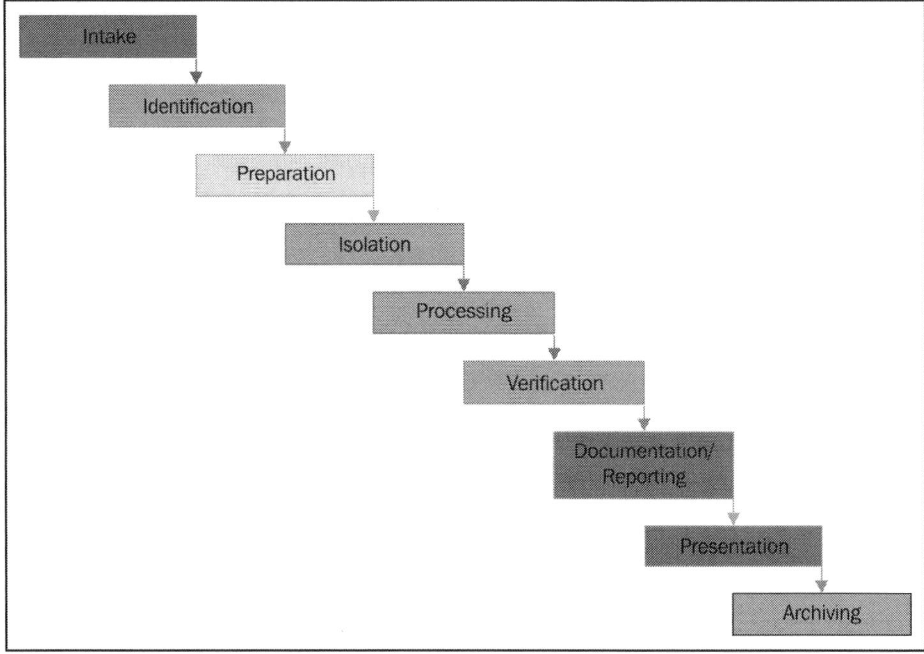

Mobile phone evidence extraction process

The evidence intake phase

The evidence intake phase is the starting phase and entails request forms and paperwork to document ownership information and the type of incident the mobile device was involved in, and outlines the type of data or information the requester is seeking. Developing specific objectives for each examination is the critical part of this phase. It serves to clarify the examiner's goals.

The identification phase

The forensic examiner should identify the following details for every examination of a mobile device:

- The legal authority
- The goals of the examination

- The make, model, and identifying information for the device
- Removable and external data storage
- Other sources of potential evidence

We will discuss each of them in the following sections.

The legal authority

It is important for the forensic examiner to determine and document what legal authority exists for the acquisition and examination of the device as well as any limitations placed on the media prior to the examination of the device.

The goals of the examination

The examiner will identify how in-depth the examination needs to be based upon the data requested. The goal of the examination makes a significant difference in selecting the tools and techniques to examine the phone and increases the efficiency of the examination process.

The make, model, and identifying information for the device

As part of the examination, identifying the make and model of the phone assists in determining what tools would work with the phone.

Removable and external data storage

Many mobile phones provide an option to extend the memory with removable storage devices, such as the Trans Flash Micro SD memory expansion card. In cases when such a card is found in a mobile phone that is submitted for examination, the card should be removed and processed using traditional digital forensic techniques. It is wise to also acquire the card while in the mobile device to ensure data stored on both the handset memory and card are linked for easier analysis. This will be discussed in detail in upcoming chapters.

Other sources of potential evidence

Mobile phones act as good sources of fingerprint and other biological evidence. Such evidence should be collected prior to the examination of the mobile phone to avoid contamination issues unless the collection method will damage the device. Examiners should wear gloves when handling the evidence.

The preparation phase

Once the mobile phone model is identified, the preparation phase involves research regarding the particular mobile phone to be examined and the appropriate methods and tools to be used for acquisition and examination.

The isolation phase

Mobile phones are by design intended to communicate via cellular phone networks, Bluetooth, Infrared, and wireless (Wi-Fi) network capabilities. When the phone is connected to a network, new data is added to the phone through incoming calls, messages, and application data, which modifies the evidence on the phone. Complete destruction of data is also possible through remote access or remote wiping commands. For this reason, isolation of the device from communication sources is important prior to the acquisition and examination of the device. Isolation of the phone can be accomplished through the use of faraday bags, which block the radio signals to or from the phone. Past research has found inconsistencies in total communication protection with faraday bags. Therefore, network isolation is advisable. This can be done by placing the phone in radio frequency shielding cloth and then placing the phone into airplane or flight mode.

The processing phase

Once the phone has been isolated from the communication networks, the actual processing of the mobile phone begins. The phone should be acquired using a tested method that is repeatable and is as forensically sound as possible. Physical acquisition is the preferred method as it extracts the raw memory data and the device is commonly powered off during the acquisition process. On most devices, the least amount of changes occur to the device during physical acquisition. If physical acquisition is not possible or fails, an attempt should be made to acquire the file system of the mobile device. A logical acquisition should always be obtained as it may contain only the parsed data and provide pointers to examine the raw memory image.

The verification phase

After processing the phone, the examiner needs to verify the accuracy of the data extracted from the phone to ensure that data is not modified. The verification of the extracted data can be accomplished in several ways.

Comparing extracted data to the handset data

Check if the data extracted from the device matches the data displayed by the device. The data extracted can be compared to the device itself or a logical report, whichever is preferred. Remember, handling the original device may make changes to the only evidence—the device itself.

Using multiple tools and comparing the results

To ensure accuracy, use multiple tools to extract the data and compare results.

Using hash values

All image files should be hashed after acquisition to ensure data remains unchanged. If file system extraction is supported, the examiner extracts the file system and then computes hashes for the extracted files. Later, any individually extracted file hash is calculated and checked against the original value to verify the integrity of it. Any discrepancy in a hash value must be explainable (for example, if the device was powered on and then acquired again, thus the hash values are different).

The document and reporting phase

The forensic examiner is required to document throughout the examination process in the form of contemporaneous notes relating to what was done during the acquisition and examination. Once the examiner completes the investigation, the results must go through some form of peer-review to ensure the data is checked and the investigation is complete. The examiner's notes and documentation may include information such as the following:

- Examination start date and time
- The physical condition of the phone
- Photos of the phone and individual components
- Phone status when received—turned on or off
- Phone make and model
- Tools used for the acquisition
- Tools used for the examination
- Data found during the examination
- Notes from peer-review

The presentation phase

Throughout the investigation, it is important to make sure that the information extracted and documented from a mobile device can be clearly presented to any other examiner or to a court. Creating a forensic report of data extracted from the mobile device during acquisition and analysis is important. This may include data in both paper and electronic formats. Your findings must be documented and presented in a manner that the evidence speaks for itself when in court. The findings should be clear, concise, and repeatable. Timeline and link analysis, features offered by many commercial mobile forensics tools, will aid in reporting and explaining findings across multiple mobile devices. These tools allow the examiner to tie together the methods behind the communication of multiple devices.

The archiving phase

Preserving the data extracted from the mobile phone is an important part of the overall process. It is also important that the data is retained in a useable format for the ongoing court process, for future reference, should the current evidence file become corrupt, and for record keeping requirements. Court cases may continue for many years before the final judgment is arrived at, and most jurisdictions require that data be retained for long periods of time for the purposes of appeals. As the field and methods advance, new methods for pulling data out of a raw, physical image may surface, and then the examiner can revisit the data by pulling a copy from the archives.

Practical mobile forensic approaches

Similar to any forensic investigation, there are several approaches that can be used for the acquisition and examination/analysis of data from mobile phones. The type of mobile device, the operating system, and the security setting generally dictate the procedure to be followed in a forensic process. Every investigation is distinct with its own circumstances, so it is not possible to design a single definitive procedural approach for all the cases. The following details outline the general approaches followed in extracting data from mobile devices.

Mobile operating systems overview

One of the major factors in the data acquisition and examination/analysis of a mobile phone is the operating system. Starting from low-end mobile phones to smartphones, mobile operating systems have come a long way with a lot of features. Mobile operating systems directly affect how the examiner can access the mobile device. For example, Android OS gives terminal-level access whereas iOS does not give such an option. A comprehensive understanding of the mobile platform helps the forensic examiner make sound forensic decisions and conduct a conclusive investigation. While there is a large range of smart mobile devices, four main operating systems dominate the market, namely, Google Android, Apple iOS, RIM BlackBerry OS, and Windows Phone. More information can be found at `http://www.idc.com/getdoc.jsp?containerId=prUS23946013`. This book covers forensic analysis of these four mobile platforms. The following is a brief overview of leading mobile operating systems.

Android

Android is a Linux-based operating system, and it's a Google open source platform for mobile phones. Android is the world's most widely used smartphone operating system. Sources show that Apple's iOS is a close second (`http://www.forbes.com/sites/tonybradley/2013/11/15/android-dominates-market-share-but-apple-makes-all-the-money/`). Android has been developed by Google as an open and free option for hardware manufacturers and phone carriers. This makes Android the software of choice for companies who require a low-cost, customizable, lightweight operating system for their smart devices without developing a new OS from scratch. Android's open nature has further encouraged the developers to build a large number of applications and upload them onto Android Market. Later, end users can download the application from Android Market, which makes Android a powerful operating system. More details on Android are covered in *Chapter 7, Understanding Android*.

iOS

iOS, formerly known as the iPhone operating system, is a mobile operating system developed and distributed solely by Apple Inc. iOS is evolving into a universal operating system for all Apple mobile devices, such as iPad, iPod touch, and iPhone. iOS is derived from OS X, with which it shares the Darwin foundation, and is therefore a Unix-like operating system. iOS manages the device hardware and provides the technologies required to implement native applications. iOS also ships with various system applications, such as Mail and Safari, which provide standard system services to the user. iOS native applications are distributed through AppStore, which is closely monitored by Apple. More details about iOS are covered in *Chapter 2, Understanding the Internals of iOS Devices.*

Windows phone

Windows phone is a proprietary mobile operating system developed by Microsoft for smartphones and pocket PCs. It is the successor to Windows mobile and primarily aimed at the consumer market rather than the enterprise market. The Windows Phone OS is similar to the Windows desktop OS, but it is optimized for devices with a small amount of storage. Windows Phone basics and forensic techniques are discussed in *Chapter 12, Windows Phone Forensics.*

BlackBerry OS

BlackBerry OS is a proprietary mobile operating system developed by BlackBerry Ltd., known as **Research in Motion (RIM)**, exclusively for its BlackBerry line of smartphones and mobile devices. BlackBerry mobiles are widely used in corporate companies and offer native support for corporate mail via MIDP, which enables wireless sync with Microsoft Exchange, e-mail, contacts, calendar, and so on, while used along with the BlackBerry Enterprise server. These devices are known for their security. BlackBerry OS basics and forensic techniques are covered in *Chapter 13, BlackBerry Forensics.*

Mobile forensic tool leveling system

Mobile phone forensic acquisition and analysis involves manual effort and the use of automated tools. There are a variety of tools that are available for performing mobile forensics. All the tools have their pros and cons, and it is fundamental that you understand that no single tool is sufficient for all purposes. So understanding the various types of mobile forensic tools is important for forensic examiners. When identifying the appropriate tools for the forensic acquisition and analysis of mobile phones, a mobile device forensic tool classification system (shown in the following figure) developed by Sam Brothers comes in handy for the examiners.

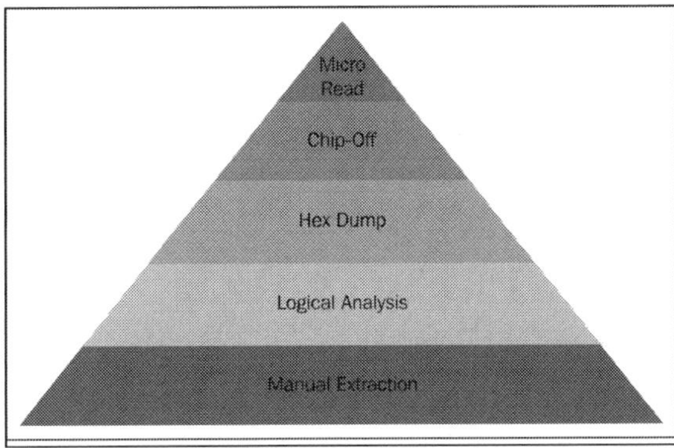

Cellular phone tool leveling pyramid (Sam Brothers, 2009)

The objective of the mobile device forensic tool classification system is to enable an examiner to categorize the forensic tools based upon the examination methodology of the tool. Starting at the bottom of the classification and working upward, the methods and the tools generally become more technical, complex, and forensically sound, and require longer analysis times. There are pros and cons of performing an analysis at each layer. The forensic examiner should be aware of these issues and should only proceed with the level of extraction that is required. Evidence can be destroyed completely if the given method or tool is not properly utilized. This risk increases as you move up in the pyramid. Thus, proper training is required to obtain the highest success rate in data extraction from mobile devices.

Each existing mobile forensic tool can be classified under one or more of the five levels. The following sections contain a detailed description of each level.

Manual extraction

This method involves simply scrolling through the data on the device and viewing the data on the phone directly through the use of the device's keypad or touchscreen. The information discovered is then photographically documented. The extraction process is fast and easy to use, and will work on almost every phone. This method is prone to human error, such as missing certain data due to unfamiliarity with the interface. At this level, it is not possible to recover deleted information and grab all the data. There are some tools that have been developed to aid an examiner to easily document a manual extraction.

Logical extraction

Logical extraction involves connecting the mobile device to forensic hardware or to a forensic workstation via a USB cable, RJ-45 cable, Infrared, or Bluetooth. Once connected, the computer initiates a command and sends it to the device, which is then interpreted by the device processor. Next, the requested data is received from the device's memory and sent back to the forensic workstation. Later, the examiner can review the data. Most of the forensic tools currently available work at this level of the classification system. The extraction process is fast, easy to use, and requires little training for the examiners. On the flip side, the process may write data to the mobile and might change the integrity of the evidence. In addition, deleted data is almost never accessible.

Hex dump

A hex dump, also referred to as a physical extraction, is achieved by connecting the device to the forensic workstation and pushing unsigned code or a bootloader into the phone and instructing the phone to dump memory from the phone to the computer. Since the resulting raw image is in binary format, technical expertise is required to analyze it. The process is inexpensive, provides more data to the examiner, and allows the recovering of the deleted files from the device-unallocated space on most devices.

Chip-off

Chip-off refers to the acquisition of data directly from the device's memory chip. At this level, the chip is physically removed from the device and a chip reader or a second phone is used to extract data stored on it. This method is more technically challenging as a wide variety of chip types are used in mobiles. The process is expensive and requires hardware level knowledge as it involves the de-soldering and heating of the memory chip. Training is required to successfully perform a chip-off extraction. Improper procedures may damage the memory chip and render all data unsalvageable. When possible, it is recommended that the other levels of extraction are attempted prior to chip-off since this method is destructive in nature. Also, the information that comes out of memory is in a raw format and has to be parsed, decoded, and interpreted. The chip-off method is preferred in situations where it is important to preserve the state of memory exactly as it exists on the device. It is also the only option when a device is damaged but the memory chip is intact.

The chips on the device are often read using the **Joint Test Action Group** (**JTAG**) method. The JTAG method involves connecting to **Test Access Ports** (**TAPs**) on a device and instructing the processor to transfer the raw data stored on memory chips. The JTAG method is generally used with devices that are operational but inaccessible using standard tools.

Micro read

The process involves manually viewing and interpreting data seen on the memory chip. The examiner uses an electron microscope and analyzes the physical gates on the chip and then translates the gate status to 0's and 1's to determine the resulting ASCII characters. The whole process is time consuming and costly, and it requires extensive knowledge and training on flash memory and the file system. Due to the extreme technicalities involved in micro read, it would be only attempted for high-profile cases equivalent to a national security crisis after all other level extraction techniques have been exhausted. The process is rarely performed and is not well documented at this time. Also, there are currently no commercial tools available to perform a micro read.

Data acquisition methods

Data acquisition is the process of imaging or otherwise extracting information from a digital device and its peripheral equipment and media. Acquiring data from a mobile phone is not as simple as a standard hard drive forensic acquisition. The following points break down the three types of forensic acquisition methods for mobile phones: **physical**, **logical**, and **manual**. These methods may have some overlap with a couple of levels discussed in the mobile forensics tool leveling system. The amount and type of data that can be collected will vary depending on the type of acquisition method being used.

Physical acquisition

Physical acquisition of mobile phones is performed using mobile forensic tools and methods. Physical extraction acquires information from the device by direct access to the flash memory. The process creates a bit-for-bit copy of an entire file system, similar to the approach taken in computer forensic investigations. A physical acquisition is able to acquire all of the data present on a device including the deleted data and access to unallocated space on most devices.

Logical acquisition

Logical acquisition of mobile phones is performed using the device manufacturer application-programming interface for synchronizing the phones contents with a computer. Many of the forensic tools perform a logical acquisition. However, the forensic analyst must understand how the acquisition occurs and whether the mobile is modified in any way during the process. Depending on the phone and forensic tools used, all or some of the data is acquired. A logical acquisition is easy to perform and only recovers the files on a mobile phone and does not recover data contained in unallocated space.

Manual acquisition

With mobile phones, physical acquisition is usually the best option, and logical acquisition is the second-best option. Manual extraction should be the last option when performing the forensic acquisition of a mobile phone. Both logical and manual acquisition can be used to validate findings in the physical data. During manual acquisition, the examiner utilizes the user interface to investigate the contents of the phone's memory. The device is used normally through a keypad or touchscreen and menu navigation, and the examiner takes pictures of each screen's contents. Manual extraction introduces a greater degree of risk in the form of human error, and there is a chance of deleting the evidence. Manual acquisition is easy to perform and only acquires the data that appears on a mobile phone.

Potential evidence stored on mobile phones

The range of information that can be obtained from mobile phones is detailed in this section. Data on a mobile phone can be found in a number of locations: SIM card, external storage card, and phone memory. In addition, the service provider also stores communication-related information. The book primarily focuses on data acquired from the phone memory. Mobile device data extraction tools recover data from the phone's memory. Even though data recovered during a forensic acquisition depends on the mobile model, in general, the data in the next set of bullet items is common across all models and useful as evidence. Note that most of the following artifacts contain date and time stamps:

- **Address Book**: This stores contact names, numbers, e-mail addresses, and so on
- **Call History**: This contains dialed, received, missed calls, and call durations
- **SMS**: This contains sent and received text messages
- **MMS**: This contains media files such as sent and received photos and videos
- **E-mail**: This contains sent, drafted, and received e-mail messages
- **Web browser history**: This contains the history of websites that were visited
- **Photos**: This contains pictures that are captured using the mobile phone camera, those downloaded from the Internet, and the ones transferred from other devices
- **Videos**: This contains videos that are captured using the mobile camera, those downloaded from the Internet, and the ones transferred from other devices

- **Music**: This contains music files downloaded from the Internet and those transferred from other devices
- **Documents**: This contains documents created using the device's applications, those downloaded from the Internet, and the ones transferred from other devices
- **Calendar**: This contains calendar entries and appointments
- **Network communication**: This contains GPS locations
- **Maps**: This contains looked-up directions, and searched and downloaded maps
- **Social networking data**: This contains data stored by applications, such as Facebook, Twitter, LinkedIn, Google+, and WhatsApp
- **Deleted data**: This contains information deleted from the phone

Rules of evidence

Courtrooms rely more and more on the information inside a mobile phone as vital evidence. Prevailing evidence in court requires a good understanding of the rules of evidence. Mobile forensics is a relatively new discipline and laws dictating the validity of evidence are not widely known. However, there are five general rules of evidence that apply to digital forensics and need to be followed in order for evidence to be useful. Ignoring these rules makes evidence inadmissible, and your case could be thrown out. These five rules are—admissible, authentic, complete, reliable, and believable.

Admissible

This is the most basic rule and a measure of evidence validity and importance. The evidence must be preserved and gathered in such a way that it can be used in court or elsewhere. Many errors can be made that could cause a judge to rule a piece of evidence as inadmissible. For example, evidence that is gathered using illegal methods is commonly ruled inadmissible.

Authentic

The evidence must be tied to the incident in a relevant way to prove something. The forensic examiner must be accountable for the origin of the evidence.

Complete

When evidence is presented, it must be clear and complete and should reflect the whole story. It is not enough to collect evidence that just shows one perspective of the incident. Presenting incomplete evidence is more dangerous than not providing any evidence at all as it could lead to a different judgment.

Reliable

Evidence collected from the device must be reliable. This depends on the tools and methodology used. The techniques used and evidence collected must not cast doubt on the authenticity of the evidence. If the examiner used some techniques that cannot be reproduced, the evidence is not considered unless they were directed to do so. This would include possible destructive methods such as chip-off extraction.

Believable

A forensic examiner must be able to explain, with clarity and conciseness, what processes they used and the way the integrity of the evidence was preserved. The evidence presented by the examiner must be clear, easy to understand, and believable by jury.

Good forensic practices

Good forensic practices apply to the collection and preservation of evidence. Following the good forensic practices ensures that evidence will be accepted in a court as being authentic and accurate. Modification of evidence, either intentionally or accidentally, can affect the case. So, understanding the best practices is critical for forensic examiners.

Securing the evidence

With advanced smartphone features such as **Find My iPhone** and remote wipes, securing a mobile phone in a way that it cannot be remotely wiped is of great importance. Also, when the phone is powered on and has service, it constantly receives new data. To secure the evidence, use the right equipment and techniques to isolate the phone from all networks. With isolation, the phone is prevented from receiving any new data that would cause active data to be deleted.

Preserving the evidence

As evidence is collected, it must be preserved in a state that is acceptable in court. Working directly on the original copies of evidence might alter it. So, as soon as you recover a raw disk image or files, create a read-only master copy and duplicate it. In order for evidence to be admissible, there must be a method to verify that the evidence presented is exactly the same as the original collected. This can be accomplished by creating a hash value of the image. After duplicating the raw disk image or files, compute and verify the hash values for the original and the copy to ensure that the integrity of the evidence is maintained. Any changes in hash values should be documented and explainable. All further processing or examination should be performed on copies of the evidence. Any use of the device might alter the information stored on the handset. So, perform only the tasks that are absolutely necessary.

Documenting the evidence

Be sure to document all the methods and tools that are used to collect and extract the evidence. Detail your notes so that another examiner could reproduce them. Your work must be reproducible; if not, a judge may rule it inadmissible.

Documenting all changes

It's important to document the entire recovery process, including all the changes made during the acquisition and examination. For example, if the forensic tool used for the data extraction sliced up the disk image to store it, this must be documented. All changes to the mobile device, including power cycling and syncing, should be documented in your case notes.

Summary

Mobile device forensics includes many approaches and concepts that fall outside of the boundaries of traditional digital forensics. Examiners responsible for mobile devices must understand the different acquisition methods and the complexities of handling the data during analysis. Extracting data from a mobile device is half the battle. The operating system, security features, and type of smartphone will determine the amount of access you have to the data. The next chapter will provide insight to iOS forensics. You will learn about the file system layout, security features, and the way the files are stored on the iOS device.

2

Understanding the Internals of iOS Devices

As of September 2013, Apple had sold more than 550 million iOS devices (170 million iPads and 387 million iPhones) according to released sales records. While iOS is the leading operating system for tablets worldwide, Android continues to be the leading operating system for smartphones worldwide. The following screenshot represents the worldwide mobile/tablet operating system share from 2013 to 2014 according to `https://www.netmarketshare.com/operating-system-market-share.aspx?qprid=9&qpcustomb=1`:

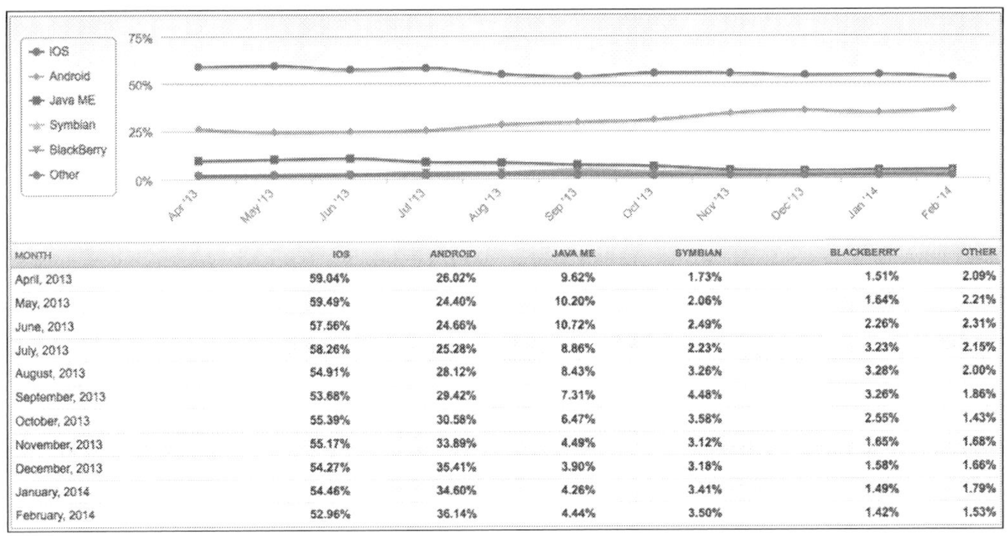

MONTH	IOS	ANDROID	JAVA ME	SYMBIAN	BLACKBERRY	OTHER
April, 2013	59.04%	26.02%	9.62%	1.73%	1.51%	2.09%
May, 2013	59.49%	24.40%	10.20%	2.06%	1.64%	2.21%
June, 2013	57.56%	24.66%	10.72%	2.49%	2.26%	2.31%
July, 2013	58.26%	25.28%	8.86%	2.23%	3.23%	2.15%
August, 2013	54.91%	28.12%	8.43%	3.26%	3.28%	2.00%
September, 2013	53.68%	29.42%	7.31%	4.48%	3.26%	1.86%
October, 2013	55.39%	30.58%	6.47%	3.58%	2.55%	1.43%
November, 2013	55.17%	33.89%	4.49%	3.12%	1.65%	1.68%
December, 2013	54.27%	35.41%	3.90%	3.18%	1.58%	1.66%
January, 2014	54.46%	34.60%	4.26%	3.41%	1.49%	1.79%
February, 2014	52.96%	36.14%	4.44%	3.50%	1.42%	1.53%

Regardless of the statistics, if you are a forensic examiner, chances are you will need to conduct an examination of an iOS mobile device.

In order to perform a forensic examination on an iOS device, the examiner must understand the internal components and inner workings of that device. Developing an understanding of the underlying components of a mobile device will help the forensic examiner understand the criticalities involved in the forensic process, including what data can be acquired, where the data is stored, and what methods can be used to access the data from that device. So, before we delve into the examination of iOS devices, it is necessary to know the different models that exist and their internals.

This book primarily focuses on the iPhone and forensic techniques associated with it. However, the same techniques may be applied to other Apple devices, such as the iPod Touch, iPad, and Apple TV.

iPhone models

The iPhone is among the most popular mobile phones on the market. Apple released the first generation iPhone in June 2007. Ever since the first release, the iPhone has gained a lot of popularity due to its advanced functionality and usability. The introduction of the iPhone has redefined the entire world of mobile computing. Consumers started looking for faster and more efficient phones. Various iPhone models exist now with different features and storage capabilities to serve the consumer requirements. The following table lists all the iPhone models and its initial iOS versions. With the iPhone, individuals can access e-mail, take photos and videos, listen to music, browse the Internet, and do much more. Furthermore, endless applications are available for download to extend the standard capabilities that exist on the iPhone.

Device	Model	Initial OS	Internal name	Identifier	Release date
iPhone 2G	A1203	iPhone OS 1.0	M68AP	iPhone 1,1	June 2007
iPhone 3G	A1241	iPhone OS 2.0	N82AP	iPhone 1,2	July 2008
iPhone 3G (china)	A1324				
iPhone 3GS	A1303	iPhone OS 3.0	N88AP	iPhone 2,1	June 2009
iPhone 3GS (china)	A1325				
iPhone 4 - GSM	A1332	iOS 4.0	N90AP	iPhone 3,1	June 2010
iPhone 4 - CDMA	A1349		N92AP	iPhone 3,2	

Device	Model	Initial OS	Internal name	Identifier	Release date
iPhone 4S	A1387	iOS 5.0	N94AP	iPhone 4,1	October 2011
iPhone 4S (China)	A1431				
iPhone 5	A1428	iOS 6.0	N41AP	iPhone 5,1	September 2012
iPhone 5 rev2	A1429		N42AP	iPhone 5,2	
	A1442				
iPhone 5C - GSM	A1456	iOS 7.0	N48AP	iPhone 5,3	
	A1532				
iPhone 5C - CDMA	A1507		N49AP	iPhone 5,4	
	A1516				
	A1526				
	A1529				
iPhone 5S - GSM	A1433	iOS 7.0	N51AP	iPhone 6,1	September 2013
	A1533				
iPhone 5S - CDMA	A1457		N53AP	iPhone 6,2	
	A1518				
	A1528				
	A1530				

iPhone models

The most recent iPhones, the seventh generation iPhone 5C and iPhone 5S, were just released at the time of writing this book. Currently, there is no method or tool available to physically recover data from these devices. However, the file system and a logical acquisition can be obtained if the iPhone is unlocked. Acquisition methods for data extraction are available and will be discussed in *Chapter 3, Data Acquisition from iOS Devices*, and *Chapter 4, Data Acquisition from iOS Backups*.

Before examining an iPhone, it is necessary to identify the correct hardware model and the firmware version installed on the device. Knowing the iPhone details helps you to understand the criticalities and possibilities of obtaining evidence from the iPhone. For example, in many cases, the device passcode is required in order to obtain the file system or logical image. Depending on the iOS version, device model, and passcode complexity, it may be possible to obtain the device passcode using a brute force attack.

There are various ways to identify the hardware of a device. The easiest way to identify the hardware of a device is by observing the **model number** displayed on the back of the device. The following image shows the model number etched on the back of the casing. Apple's knowledge base articles can be helpful for this purpose. Details on identifying iPhone models can be found at `http://support.apple.com/kb/HT3939`.

iPhone model number located on the back of the case

The firmware version of an iPhone can be found by accessing the **Settings** option and then navigating to **General** | **About** | **Version**, as shown in the following screenshot. The purpose of the firmware is to enable certain features and assist with the general functioning of the device.

Airtel 14:35 35 %

General About

Network	AirTel
Songs	15
Videos	0
Photos	173
Applications	26
Capacity	13.7 GB
Available	10.8 GB
Version	5.1.1 (9B206)
Carrier	AirTel 12.0

The iPhone About screen, displaying firmware Version 5.1.1 (9B206)

Alternatively, the **ideviceinfo** command-line tool available in the `libimobiledevice` software library (`http://www.libimobiledevice.org/`) can be used to identify the iPhone model and its iOS version. The library allows you to communicate with an iPhone even if the device is locked by a passcode. The software library was developed by Nikias Bassen (pimskeks), and it was compiled for Mac OS X by Ben Clayton (benvium).

Mac OS X can be installed in virtual machines for use on a Windows platform. To obtain the iPhone model and its iOS version information on Mac OS X 10.8, the following steps must be followed:

1. Open the terminal application.
2. From the command line, run the following command to download the `libimobiledevice` library:

   ```
   $ git clone https://github.com/benvium/libimobiledevice-
   macosx.git ~/Desktop/libimobiledevice-macosx/
   ```

   The command creates the `libimobiledevice-macosx` directory on the user's desktop and places the `libimobiledevice` command-line tools onto it.

3. Navigate to the `libimobiledevice-macosx` directory, as follows:

   ```
   $ cd ~/Desktop/libimobiledevice-macosx/
   ```

4. Create and edit the `.bash_profile` file using the `nano` command, as follows:

   ```
   $ nano ~/.bash_profile
   ```

5. Add the following two lines to the `.bash_profile` file, as follows:

   ```
   export DYLD_LIBRARY_PATH=~/Desktop/libimobiledevice-
   macosx/:$DYLD_LIBRARY_PATH
   PATH=${PATH}:~/Desktop/libimobiledevice-macosx/
   ```

   Press *Ctrl* + *X*, type the letter `y` and hit *Enter* to save the file.

6. Return to the terminal and run the following command:

   ```
   $ source ~/.bash_profile
   ```

7. Connect the iPhone to the Mac workstation using a USB cable, and run the `ideviceinfo` command with the `-s` option:

   ```
   $ ./ideviceinfo -s
   ```

   Output of the `ideviceinfo` command displays the iPhone identifier, internal name, and the iOS version as shown:

   ```
   BuildVersion: 9B206
   ```

```
DeviceClass: iPhone
DeviceName: iPhone4
HardwareModel: N90AP
ProductVersion: 5.1.1
ProductionSOC: true
ProtocolVersion: 2
TelephonyCapability: true
UniqueChipID: 1937316564364
WiFiAddress: 58:1f:aa:22:d1:0a
```

Every release of the iPhone comes with improved or newly added features. The following tables show the specifications and features of legacy and current iPhone models:

Specification	iPhone	iPhone 3G	iPhone 3GS
System on chip	Samsung Chip	Samsung Chip	Samsung Chip
CPU	620 MHz Samsung 32-bit RISC ARM	620 MHz Samsung 32-bit RISC ARM	833 MHz ARM Cortex-A8
Onboard RAM	128 MB	128 MB	256 MB
Screen size (in inches)	3.5	3.5	3.5
Resolution	480*320	480*320	480*320
Connectivity	Wi-Fi, Bluetooth 2.0, GSM	Wi-Fi, Bluetooth 2.0, GSM/UMTS/ HSDPA, GPS	Wi-Fi, Bluetooth 2.1, GSM, UMTS/HSDPA, GPS
Camera (megapixel)	2	2	3
Front camera	N/A	N/A	N/A
Storage (GB)	4, 8, 16	8, 16	8, 16, 32
Weight (in ounces)	4.8	4.7	4.8
Dimensions	4.5 * 2.4 * 0.46	4.55 * 2.44 * 0.48	4.55 * 2.44 * 0.48

Specification	iPhone	iPhone 3G	iPhone 3GS
Battery life			
Talk/video/web/audio	8/7/6/24	5/7/5/24	5/10/5/30
Standby time (hours)	250	300	300
Colors	Black	Black, white (white not in 8 GB)	Black, white (white not in 8 GB)
Material	Aluminum, glass, and steel	Glass, plastic, and steel	Glass, plastic, and steel
Connector	USB 2.0 dock connector	USB 2.0 dock connector	USB 2.0 dock connector
SIM card form-factor	Mini SIM	Mini SIM	Mini SIM
Siri support	No	No	No

The most recent iPhone features are shown in the following table:

Specification	iPhone 4	iPhone 4S	iPhone 5	iPhone 5C	iPhone 5S
System on chip	Apple A4	Apple A5	Apple A6	Apple A6	Apple A7
CPU	1 GHz ARM Cortex-A8	800 MHz dual core ARM Cortex-A9	1.3 GHz dual core Apple-designed ARMv7s	1.3 GHz dual core Apple-designed ARMv7s	1.3 GHz dual core Apple-designed ARMv8-A
Onboard RAM	512 MB	512 MB	1 GB	1 GB	1 GB
Screen size (in inches)	3.5	3.5	4	4	4
Resolution	960*640	960*640	1136*640	1136*640	1136*640

Specification	iPhone 4	iPhone 4S	iPhone 5	iPhone 5C	iPhone 5S
Connectivity	Wi-Fi, Bluetooth 2.1, GSM, UMTS/ HSDPA/ HSUPA, GPS	Wi-Fi, Bluetooth 4, GSM, UMTS/ HSDPA/ HSUPA, GPS	Wi-Fi, Bluetooth 4, UMTS/ HSDPA+/ DC- HSDPA, GSM, GPS	Wi-Fi, Bluetooth 4, UMTS/ HSDPA+/ DC-HSDPA/ LTE, GSM, GPS	Wi-Fi, Bluetooth 4, UMTS/ HSDPA+/DC- HSDPA/LTE/ TD-LTE, GSM, GPS
Camera (megapixel)	5	8	8	8	8
Front camera	VGA	VGA	720P	720P	720P
Storage (GB)	8, 16, 32	8, 16, 32, 64	16, 32, 64	8, 16, 32, 64	8, 16, 32, 64
Weight (in ounces)	4.8	4.9	3.95	4.7	4
Dimensions	4.5 * 2.31 * 0.37	4.5 * 2.31 * 0.37	4.87 * 2.31 * 0.30	4.98 * 2.33 * 0.353	4.87 * 2.31 * 0.3
Battery life Talk/video/ web/audio	7/10/10/40	8/10/9/40	8/10/10/40	10/10/10/40	10/10/10/40
Standby time (hours)	300	300	225	250	250
Colors	Black	Black, white	Black, white	White, pink, yellow, blue, or green	Silver, space gray, or gold
Material	Aluminosilicate glass and stainless steel	Aluminosilicate glass and stainless steel	Black - anodized aluminum slate metal white - silver aluminum metal	White, pink, yellow, blue, or green	White or black front with aluminum metal back
Connector	USB 2.0 dock connector	USB 2.0 dock connector	Lightning connector	Lightning connector	Lightning connector
SIM card form-factor	Micro SIM	Micro SIM	Nano-SIM	Nano-SIM	Nano-SIM
Siri support	No	Yes	Yes	Yes	Yes

One of the major changes in the iPhone 5, iPhone 5C, and iPhone 5S is the USB dock connector, which is used to charge and synchronize the device with the computer. Devices prior to the iPhone 5 use a 30-pin USB dock connector, whereas the newer iPhones use an eight-pin lightning connector.

iPhone hardware

The iPhone is a collection of modules, chips, and electronic components from different manufacturers. Due to the complexities of the iPhone, the list of hardware components is extensive. A detailed list of iPhone hardware components is defined at `https://viaforensics.com/resources/white-papers/iphone-forensics/overview`.

The following images show the internals of the iPhone 5S. The images were taken after dismantling the iPhone 5S. Internal images for all iPhones can be found in the teardown section from `http://www.ifixit.com/Device/iPhone`.

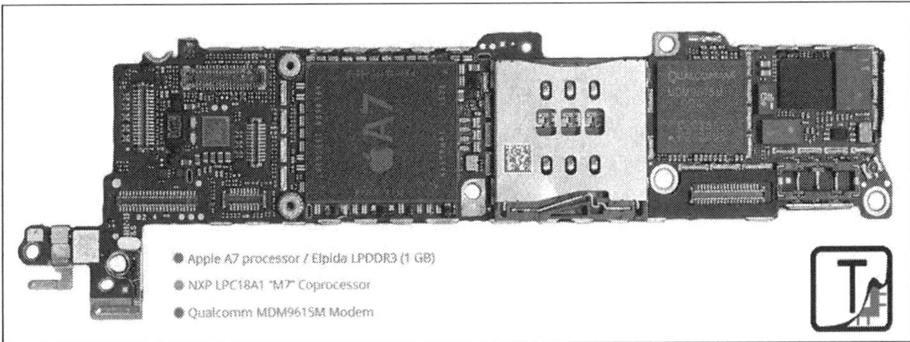

The iPhone 5S teardown image—side one (included with kind permission from TechInsights)

And the following is the image showing the back of the iPhone 5S:

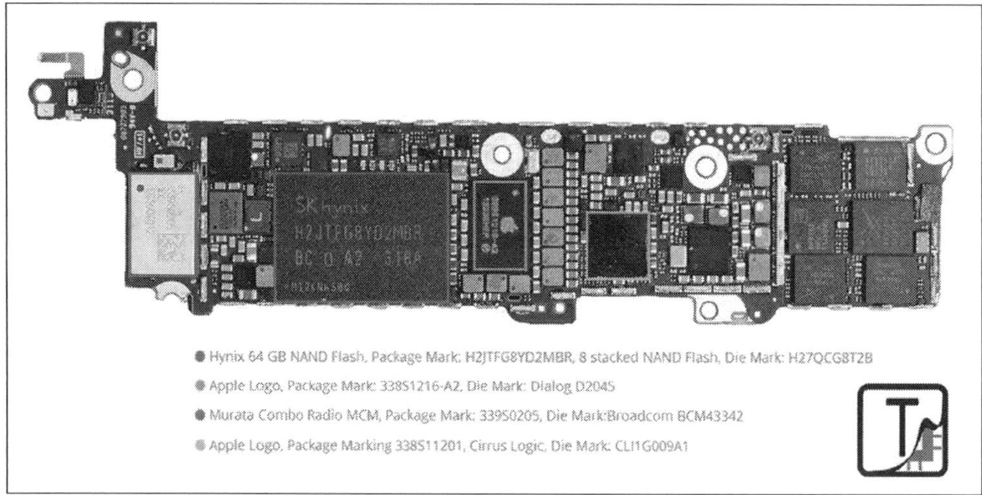

The iPhone 5S teardown image—side two (included with kind permission from TechInsights)

iPad models

The Apple iPhone changed the way cell phones are produced and used. Similarly, the iPad, a version of the tablet computer introduced in January 2010, squashed the sales of notebooks. With the iPad, individuals can shoot video, take photos, play music, read books, browse the Internet, and do much more. Various iPad models exist now with different features and storage capabilities. The following table lists all the iPad models and their initial iOS versions. Details on identifying iPad models can be found at `http://support.apple.com/kb/ht5452`.

Device	Model	Initial OS	Internal name	Identifier	Release date
iPad - Wi-Fi	A1219	iOS 3.2	K48AP	iPad 1,1	January 2010
iPad - 3G	A1337			iPad 1,1	
iPad 2 - Wi-Fi	A1395	iOS 4.3	K93AP	iPad 2,1	March 2011
iPad 2 - GSM	A1396		K94AP	iPad 2,2	
iPad 2 - CDMA	A1397		K95AP	iPad 2,3	

Device	Model	Initial OS	Internal name	Identifier	Release date
iPad 2 - Wi-Fi rev	A1395		K93AAP	iPad 2,4	
iPad 3 - Wi-Fi	A1416		J1AP	iPad 3,1	
iPad 3 - Wi-Fi + Cellular Verizon	A1403	iOS 5.1	J2AP	iPad 3,2	March 2012
iPad 3 - Wi-Fi + Cellular AT&T	A1430		J2AAP	iPad 3,3	
iPad 4 - Wi-Fi	A1458	iOS 6.0	P101AP	iPad 3,4	
iPad 4 -Wi-Fi + Cellular AT &T	A1459		P102AP	iPad 3,5	
iPad 4 - Wi-Fi + Cellular Verizon	A1460	iOS 6.0.1	P103AP	iPad 3,6	
iPad mini - Wi-Fi	A1432	iOS 6.0	P105AP	iPad 2,5	October 2012
iPad mini -Wi-Fi + Cellular AT&T	A1454		P106AP	iPad 2,6	
iPad mini - Wi-Fi + Cellular Verizon and Sprint	A1455	iOS 6.0.1	P107AP	iPad 2,7	
iPad Air - Wi-Fi	A1474	iOS 7.0.3	J71AP	iPad 4,1	November 2013
iPad Air - Wi-Fi + Cellular	A1475		J72AP	iPad 4,2	

Every release of the iPad comes with improved or newly added features. The following table shows the specifications and features of legacy and current iPad Wi-Fi models:

Specification	iPad	iPad 2	iPad 3	iPad 4	iPad Mini	iPad Air
System on chip	Apple A4	Apple A5	Apple A5X	Apple A6X	Apple A5	Apple A7
CPU	1GHz dual core Samsung-Intrinsity	1 GHz dual core ARM Cortex-A9	1 GHz dual core ARM Cortex-A9	1.4 GHz dual core Apple Swift	1 GHz dual core ARM Cortex-A9	1.4 GHz dual core ARMv8-A
Onboard RAM	256 MB	512 MB	1 GB	1 GB	512 MB	1 GB
Screen size (in inches)	9.7	9.7	9.7	9.7	7.9	9.7
Resolution	1024*768	1024*768	2048*1536	2048*1536	1024*768	2048*1536
Connectivity	Wi-Fi, Bluetooth 2.1	Wi-Fi, Bluetooth 2.1	Wi-Fi, Bluetooth 4	Wi-Fi, Bluetooth 4	Wi-Fi, Bluetooth 4	Wi-Fi, Bluetooth 4
Camera (megapixel)	N/A	0.7	5	5	5	5
Front camera	N/A	0.3 MP	0.3 MP	1.2 MP	1.2 MP	1.2 MP
Storage (GB)	16, 32, 64	16, 32, 64	16, 32, 64	16, 32, 64, 128	16, 32, 64	16, 32, 64, 128
Weight (in ounces)	24	21.6	22.9	22.9	10.8	16
Dimensions	9.56 * 7.47 * 0.5	9.5 * 7.31 * 0.34	9.5 * 7.31 * 0.37	9.5 * 7.31 * 0.37	7.87 * 5.3 * 0.28	9.4 * 6.6 * 0.29
Battery life Video/web/ audio	10/10/140	10/10/140	10/10/140	10/10/140	10/10/140	10/10/140
Standby time (hours)	1 month	1 month	1 month	1 month	1 month	1 month
Connector	USB 2.0 dock connector	USB 2.0 dock connector	USB 2.0 dock connector	Lightning connector	Lightning connector	Lightning connector

iPad hardware

One of the key successes of Apple iOS devices is the proper selection of its hardware components. Just like the iPhone, the iPad is also a collection of modules, chips, and electronic components from different manufacturers. Internal images for all iPads can be found in the teardown section of `http://www.ifixit.com/Device/iPad`.

The following images show the internals of the iPad 3. The images were taken after dismantling the iPad 3 cellular model and were obtained from `http://www.chipworks.com/`.

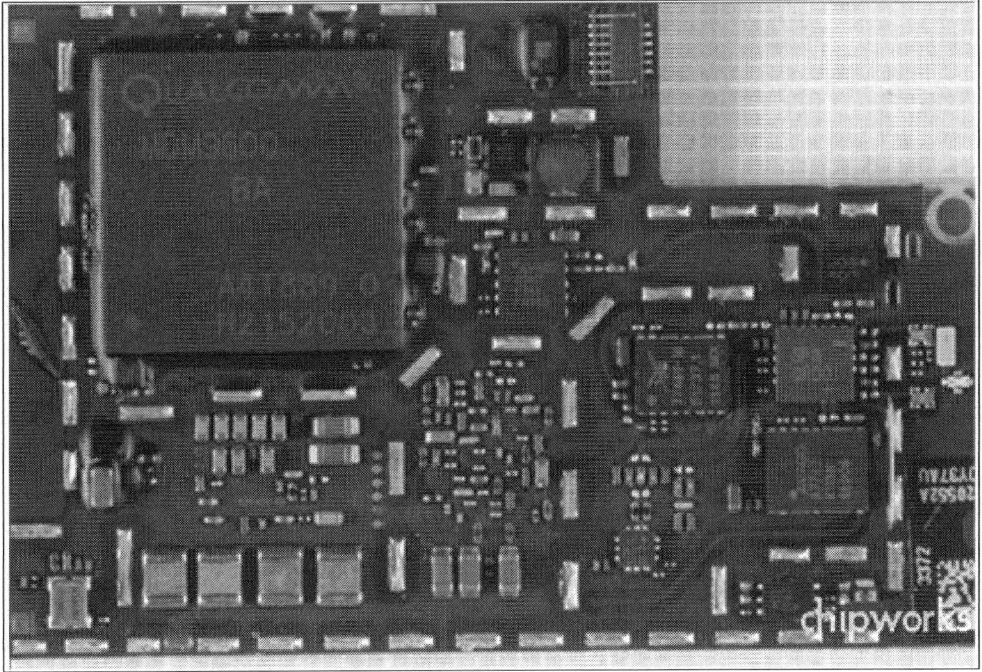

The iPad 3 cellular model teardown image—side one (included with kind permission from Chipworks)

The following image shows side two of the iPad 3 cellular model:

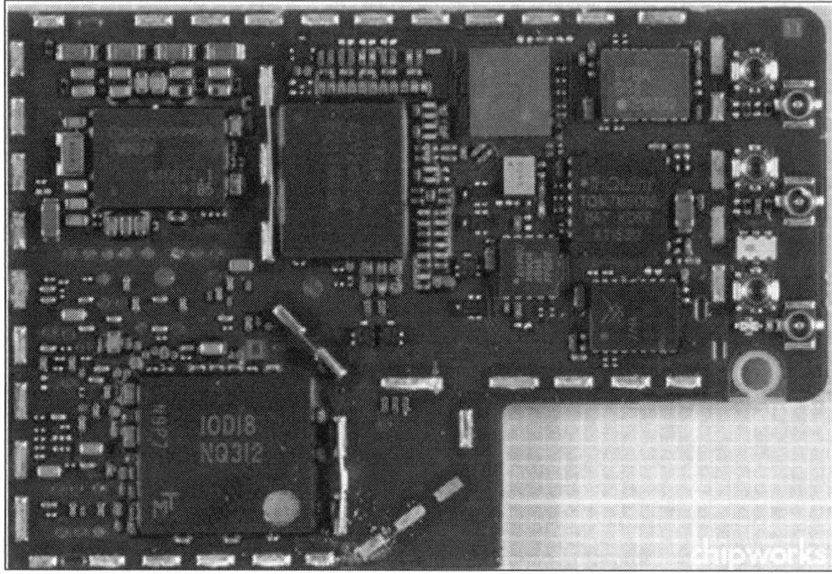

Included with kind permission from Chipworks

File system

To better understand the forensic process of an iPhone, it is good to know about the file system that is used. The file system used in the iPhone and other Apple iOS devices is **HFSX**, a variation of **HFS Plus** with one major difference. HFSX is case sensitive whereas HFS Plus is case insensitive. Other differences will be discussed later in this chapter. OS X uses HFS Plus by default and iOS uses HFSX.

The HFS Plus file system

In 1996, Apple developed a new file system, **Hierarchical File System** (**HFS**), to accommodate the storage of large datasets. In an HFS file system the storage medium is represented as volumes. HFS volumes are divided into logical blocks of 512 bytes. The logical blocks are numbered from first to last on a given volume and will remain static with the same size as physical blocks, that is, 512 bytes. These logical blocks are grouped together into allocation blocks, which are used by the HFS file system to track data in a more efficient way. HFS uses a 16-bit value to address allocation blocks, which limits the number of allocation blocks to 65,535. To overcome the inefficient allocations of disk space and some of the limitations of HFS, Apple introduced the **HFS Plus** file system (`http://dubeiko.com/development/FileSystems/HFSPLUS/tn1150.html`).

The HFS Plus file system was designed to support larger file sizes. HFS volumes are divided into sectors that are usually 512 bytes in size. These sectors are grouped together into allocation blocks. The number of allocation blocks depends on the total size of the volume. HFS Plus uses block addresses of 32 bits to address allocation blocks. HFS Plus uses journaling by default. Journaling is the process of logging every transaction to the disk, which helps in preventing file system corruption. The key characteristics of the HFS Plus file system are: efficient use of disk space, unicode support for filenames, support for name forks, file compression, journaling, dynamic resizing, dynamic defragmentation, and an ability to boot on operating systems other than Mac OS.

The HFS Plus volume

The HFS Plus volume contains a number of internal structures to manage the organization of data. These structures include a header, alternate header, and five special files: an allocation file, an Extents Overflow file, a Catalog file, an Attributes file, and a Startup file. Among the five files, three files, the Extents Overflow file, the Catalog file, and the Attribute file, use a B-tree structure, a data structure that allows data to be efficiently searched, viewed, modified, or removed. The HFS Plus volume structure is shown in the following figure:

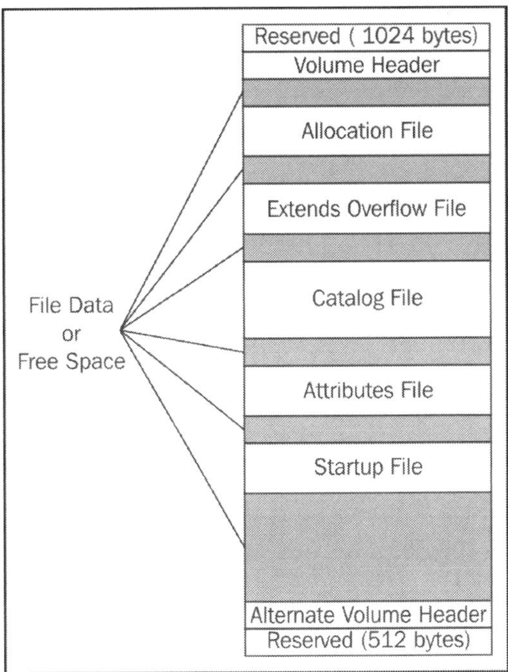

The volume structure is described as follows:

- The first 1,024 bytes are reserved for boot load information.
- **Volume Header**: This stores volume information, such as the size of allocation blocks, a timestamp of when the volume was created, and metadata about each of the five special files.
- **Allocation File**: This file is used to track which allocation blocks are in use by the system. The file format consists of one bit for every allocation block. If the bit is set, the block is in use. If it is not set, the block is free.
- **Extents Overflow File**: This file records the allocation blocks that are allocated when the file size exceeds eight blocks, which helps in locating the actual data when referred. Bad blocks are also recorded in the file.
- **Catalog File**: This file contains information about the hierarchy of files and folders, which is used to locate any file and folder within the volume.
- **Attribute File**: This file contains inline data attribute records, fork data attribute records, and extension attribute records.
- **Startup File**: This file holds the information needed to assist in booting a system that does not have HFS Plus support.
- **Alternate Volume Header**: This is a backup of the volume header, and it is primarily used for disk repair.
- The last 512 bytes are reserved for use by Apple, and it is used during the manufacturing process.

Disk layout

By default, the file system is configured as two logical disk partitions: system (root or firmware) partition and user data partition.

The system partition contains the OS and all of the preloaded applications used with the iPhone. The system partition is mounted as read-only unless an OS upgrade is performed or the device is jailbroken. The partition is updated only when a firmware upgrade is performed on the device. During this process, the entire partition is formatted by iTunes without affecting any of the user data. The system partition takes only a small portion of storage space, normally between 0.9 GB and 2.7 GB, depending on the size of the NAND drive. As the system partition was designed to remain in factory state for the entire life of the iPhone, there is typically little useful evidentiary information that can be obtained from it. If the iOS device was jailbroken, files containing information regarding the jailbreak may be resident on the system partition. Jailbreaking an iOS device allows the user root access to the device and voids the manufacturer warranty. Jailbreaking will be discussed later in this chapter.

The user data partition contains all user-created data ranging from music to contacts. The user data partition occupies most of the NAND memory and is mounted at /private/var on the device. Most of the evidentiary information can be found in this partition. During a physical acquisition, both the user data and system partitions can be captured and saved as a .dmg or .img file. These raw image files can be mounted as read-only for forensic analysis, which is covered in detail in *Chapter 3, Data Acquisition from iOS Devices*. Even on non-jailbroken iOS devices, it is recommended to acquire both the system and user data partitions to ensure all data is obtained for examination.

To view the mounted partitions on the iPhone, connect a jailbroken iPhone to a workstation over SSH, and run the `mount` command. For this example, iPhone 4 with 5.1.1 is used.

The `mount` command shows that the system partition is mounted on `/ (root)`, and the user data partition is mounted on `/private/var`, as shown in the following command lines. Both partitions show HFS as the file system, and the user data partition even shows that journaling is enabled.

```
iPhone4:~ root# mount
/dev/disk0s1s1 on / (hfs, local, journaled, noatime)
devfs on /dev (devfs, local, nobrowse)
/dev/disk0s1s2 on /private/var (hfs, local, journaled, noatime,
protect)
```

To view the raw disk images on the iPhone, connect a jailbroken iPhone to a workstation over SSH, and run the `ls -lh rdisk*` command. `rdisk0` is the entire file system and `rdisk0s1` is the firmware partition. `rdisk0s1s1` is the root file system and `rdisk0s1s2` is the user file system, as shown in the following command lines:

```
iPhone4:/dev root# ls -lh rdisk*
crw-r----- 1 root operator 14, 0 Oct 10 04:28 rdisk0
crw-r----- 1 root operator 14, 1 Oct 10 04:28 rdisk0s1
crw-r----- 1 root operator 14, 2 Oct 10 04:28 rdisk0s1s1
crw-r----- 1 root operator 14, 3 Oct 10 04:28 rdisk0s1s2
```

iPhone operating system

iOS is Apple's most advanced and feature-rich proprietary mobile operating system. It was released with the first generation of the iPhone. When introduced, it was named **iPhone OS**, and later it was renamed to **iOS** to reflect the unified nature of the operating system that powers all Apple iOS devices, such as the iPhone, iPod Touch, iPad, and Apple TV. iOS is derived from core OS X technologies and streamlined to be compact and efficient for mobile devices.

It utilizes a multitouch interface where simple gestures are used to operate and control the device, such as swiping your finger across the screen to move to the successive page or pinching your fingers to zoom. In simple terms, iOS assists with the general functioning of the device. iOS is really Mac OS X with some significant differences:

- The architecture for which the kernel and binaries are compiled is ARM-based rather than Intel x86_64
- The OS X kernel is open source, whereas the iOS kernel remains closed
- Memory management is much tighter
- The system is hardened and does not allow access to the underlying APIs

iOS history

iOS, like any other operating system, has gone through multiple updates since its release. Apple occasionally releases newer versions to enable new features, to support latest hardware, and to fix bugs. The latest version of iOS at the time of this writing is iOS 7.0.3. Though Apple sticks with a numeric approach for new iOS builds, all iOS versions have code names that are private to Apple. The following sections describe the history of iOS development.

1.x – the first iPhone

iPhone OS 1.x was the first release of Apple's touch-centric mobile operating system. On its initial release, Apple stated that the iPhone uses a version of the desktop operating system, OS X. Later it was named iPhone OS. The original build was known as **Alpine**, but the final released version was **Heavenly**.

2.x – App Store and 3G

iPhone OS 2.0 (known as **BigBear**) was released along with iPhone 3G. Features required for corporate needs such as VPN and Microsoft Exchange were introduced with this release. The big addition to the OS with this release was the **App Store**, a marketplace for the third-party applications that could run on the iPhone. Apple also released the **iPhone Software Development Kit** (**SDK**) to assist developers in creating applications on the App Store for free or for purchase. **Global Positioning System** (**GPS**) was also added to the iPhone with this release.

3.x – the first iPad

iPhone OS 3.0 (known as **Kirkwood**) became available with the release of iPhone 3GS. The iOS release brought the copy/paste feature, spotlight searches, and push notifications for third-party applications, and many other enhancements to the built-in applications. Multitasking was introduced, but it was limited to a selection of the applications Apple included on the device. The first iPad was introduced with iPhone OS 3.2 (known as **Wildcat**) and later updated to 3.2.2, a version specifically made for the iPad.

4.x – Game Center and multitasking

iOS 4.0 (known as **Apex**) was the first major release after renaming the iPhone OS to iOS. This release brought over 100 new features, such as FaceTime, iBooks, voice control, and 1,500 new APIs to the developers. Starting with this release, multitasking was extended to third-party iOS applications. Apple also released **Game Center**, an online multiplayer social gaming network along with this release.

5.x – Siri and iCloud

iOS 5.0 (known as **Telluride**) was released with iPhone 4S. iOS 5 with iPhone 4S introduced Apple's natural language-based voice control, **Siri**—a virtual assistant. This update brought many new features, such as notification center, iMessages, Newsstand, Twitter integration, the Reminders application, and **over the air (OTA)** software updates. The biggest addition to the release was the **iCloud**, Apple's cloud-based service that allows users to synchronize their contacts, calendar, pictures, and much more to the cloud.

6.x – Apple Maps

iOS 6.0 (known as **Sundance**) was released in June 2012 with the release of iPhone 5. With iOS 6, the old, Google-powered Maps application was removed, and an all-new Apple Maps with data supplied by TomTom was added. The YouTube application was also removed in this update. iOS 6 brought many new features, such as Facebook integration, FaceTime over cellular network, Passbook, and many enhancements to the built-in applications. Better privacy controls were added with this release.

7.x – the iPhone 5S and beyond

iOS 7.0 (known as **Innsbruck**) was released in September 2013 with the release of iPhone 5S. The biggest change in iOS 7 and the most important was the system-wide redesign. With this release, Apple took the interface experience from static to dynamic. A ton of new features were introduced, such as control center, Airdrop, iTunes Radio, FaceTime audio, automatic updates for applications, activation lock, and many more. With iPhone 5S, Apple's Touch ID fingerprint identity sensor, a biometric authentication technology, was introduced.

All the iOS versions are not supported by all the iOS devices. Each iOS version is compatible only with a few devices, as shown in the following iOS compatibility matrix. This table was created using `http://iossupportmatrix.com/`.The blocks in green signify that an iOS version was supported for that device. If a version is listed, it is the earliest version supported for that device. The blocks in red mean no support for that device, and the blocks in blue are still iOS versions supported by Apple.

	iPhone OS 1.0	iPhone SDK 2.0	iPhone SDK 3.0	iPhone SDK 4.0	iOS 5	iOS 6	iOS 7
iPhone	1.0		3.1.3				
iPod Touch	1.1		3.1.3				
iPhone 3G		2.0		4.2.1			
iPod Touch (2nd Gen)		2.1.1		4.2.1			
iPhone 3GS			3.0			6.1.3	
iPod Touch (3rd Gen)			3.1.1		5.1.1		
iPad (1st Gen)			3.2	4.3.5	5.1.1		
iPhone4				4.0 (GSM)/4.2.6 (CDMA)			7.0
iPod Touch (4th Gen)				4.2.1		6.1.3	
iPad2				4.3.5			7.0
iPhone 4S					5.0		7.0
iPad					5.1		7.0
iPod Touch (5th Gen)						6.0	7.0
iPad Mini						6.0	7.0
iPhone 5						6.0	7.0
iPhone 5C							7.0.1
iPhone 5S							7.0.1

The OS compatibility matrix

The iOS architecture

iOS acts as an intermediary between the underlying hardware components and the applications that appear on the screen. The applications do not talk to the underlying hardware directly. Instead, they communicate through a well-defined system interface that protects the applications from hardware changes. This abstraction makes it easy to build applications that work on devices with different hardware capability.

The iOS architecture consists of four layers: the Cocoa Touch layer, Media layer, Core Services layer, and Core OS layer, as shown in the following figure. Each layer consists of several frameworks that would help to build an application.

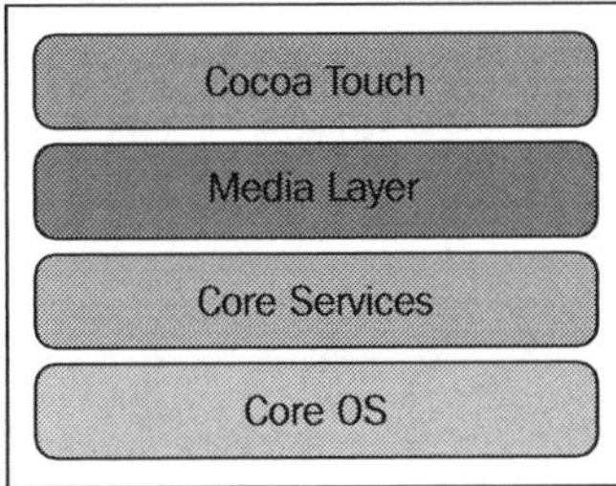

The iOS layers

The Cocoa Touch layer

The Cocoa Touch layer contains the key frameworks required to develop the visual interface for iOS applications. Frameworks in this layer provide the basic application infrastructure and support key technologies, such as multitasking and touch-based input, and many high-level system services.

The Media layer

The Media layer provides the graphics and audio and video frameworks to create the best multimedia experience available on a mobile device. The technologies in this layer help developers to build applications that look and sound great.

The Core Services layer

This Core Services layer provides the fundamental system services that are required for the applications. All these services are not used by the developers though many parts of the system are built on top of them. The layer contains the technologies to support features such as location, iCloud, and social media.

The Core OS layer

The Core OS layer is the base layer and sits directly on top of the device hardware. This layer deals with low-level functionalities and provides services such as networking (BSD sockets), memory management, threading (POSIX threads), file system handling, external accessories access, and inter-process communication.

iOS security

iOS was designed with security at its core. At the highest level, the iOS security architecture appears as shown in the following figure:

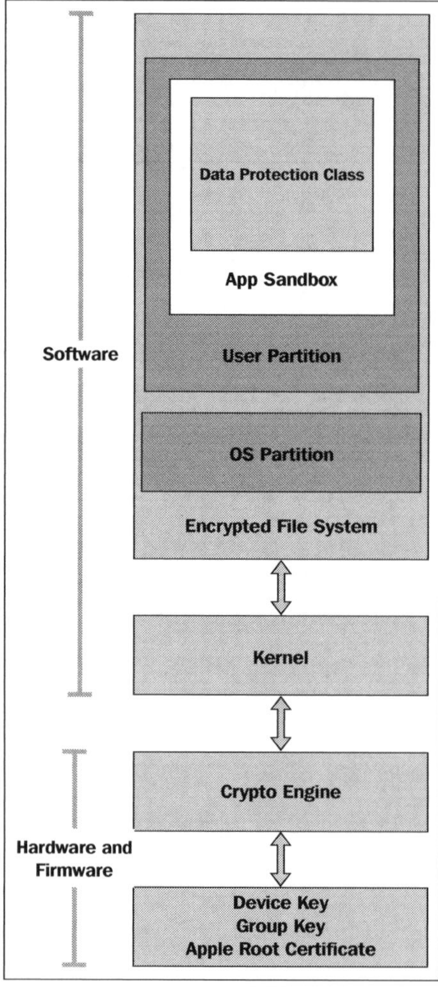

The iOS security architecture

Apple iOS devices such as iPhone, iPad, and iPod Touch are designed with layers of security. Low-level hardware features safeguard from malware attacks and the high-level OS features prevent unauthorized use. A brief overview of the iOS security features are provided in the following sections.

Passcode

Passcodes restrict unauthorized access to the device. Once a passcode is set, each time you turn on or wake up the device, it will ask for the passcode to access the device. iPhone supports simple as well as complex passcodes. iPhone 5S also supports touch ID fingerprints as a passcode.

Code signing

Code signing prevents users from downloading and installing unauthorized applications on the device. Apple says "Code Signing is the process by which your compiled iOS application is sealed and identified as yours. Also, iOS devices won't run an application or load a library unless it is signed by a trusted party. To ensure that all apps come from a known and approved source and have not been tampered with, iOS requires that all executable code be signed using an Apple-issued certificate."

Sandboxing

Sandboxing mitigates the post-code-execution exploitation by placing the application into a tightly restricted area. Applications installed on the iOS device are sandboxed, and one application cannot access the data stored by the other application.

Encryption

On iOS devices, the entire file system is encrypted with a file system key, which is computed from the device's unique hardware key.

Data protection

Data protection is designed to protect data at rest and to make offline attacks difficult. It allows applications to leverage the user's device passcode in concert with the device hardware encryption to generate a strong encryption key. Later, the strong encryption key is used to encrypt the data stored on the disk. This key prevents data from being accessed when the device is locked, ensuring that critical information is secured even if the device is compromised.

Address Space Layout Randomization

Address Space Layout Randomization (ASLR) is an exploit mitigation technique introduced with iOS 4.3. ASLR randomizes the application objects' location in the memory, making it difficult to exploit the memory corruption vulnerabilities.

Privilege separation

iOS runs with the principle of least privileges. It contains two user roles: **root** and **mobile**. The most important processes in the system run with root user privileges. All other applications that the user has direct access to, such as the browser and third-party applications, run with mobile user privileges.

Stack smashing protection

Stack smashing protection is an exploit mitigation technique. It protects against buffer overflow attacks by placing a random and known value (called **stack canary**) between a buffer and control data on the stack.

Data execution prevention

Data execution prevention (DEP) is an exploit mitigation technique mechanism in which a processor can distinguish the portions of memory that are executable code from data.

Data wipe

iOS provides an option **Erase All Content and Settings** to wipe the data on the iPhone. This type of data wipe erases user settings and information by removing the encryption keys that protects the data. As the encryption keys are erased from the device, it is not possible to recover the deleted data in forensic investigations. Other wiping methods are available that overwrite the data in the device memory. More information on wiping can be found at `http://support.apple.com/kb/ht2110`.

Activation Lock

Activation Lock, introduced with iOS 7, is a theft deterrent that works by leveraging **Find My iPhone**. When Find My iPhone is enabled, it enables the Activation Lock, and your Apple ID and password will be required to turn off Find My iPhone, to erase your device, and to reactive your device.

App Store

The App Store is an application distribution platform for iOS, developed and maintained by Apple. It is a centralized online store where users can browse and download both free and paid apps. These apps expand the functionality of a mobile device. As of December 2013, there are more than 1 million applications in the App Store, and users have downloaded them over 60 billion times. Apps available in the App Store are generally written by third-party developers. Developers use XCode and the iPhone SDK to develop iOS applications. Later, they submit the app to Apple for approval. Apple follows an extensive review process to check the app against the company guidelines. If Apple approves the app, it is published to the App Store where users can download or buy it. The strict review process makes the App Store less prone to malware. Currently, users can access the App Store via iTunes and also from their iOS devices.

Jailbreaking

Jailbreaking is the process of removing limitations imposed by Apple's mobile operating system through the use of software and hardware exploits. Jailbreaking permits unsigned code to run and gain root access on the operating system. The most common reason for jailbreaking is to expand the limited feature set imposed by Apple's App Store and to install unapproved apps. Many publicly available jailbreaking tools add an unofficial application installer to the device, such as **Cydia**, which allows users to install many third-party applications, tools, tweaks, and apps from an online file repository. The software downloaded from Cydia opens up endless possibilities on a device that a non-jailbroken device would never be able to do. The most popular jailbreaking tools are redsn0w, sn0wbreeze, evasi0n, Absinthe, seas0npass, and so on. Not all the iOS versions are jailbreakable. The website `http://www.guidemyjailbreak.com/choose-iphone-to-jailbreak/` can be helpful to find out whether a particular iOS version is jailbreakable or not and with which method. In October 2012, The U.S. Copyright Office declared that jailbreaking the iPad is illegal, while jailbreaking the iPhone is deemed legal. The governing law is reviewed every three years.

Summary

The first step in a forensic examination of an iOS device should be identifying the device model. The model of an iOS device can be used to help the examiner develop an understanding of the underlying components and capabilities of the device, which can be used to drive the methods for acquisition and examination. Legacy iOS devices should not be disregarded because they may surface as part of an investigation. Examiners must be aware of all iOS devices as old devices are sometimes still in use and may be tied to a criminal investigation. The next chapter will provide tips and techniques for acquiring data from the iOS devices discussed in this chapter.

3

Data Acquisition from iOS Devices

An iPhone recovered from a crime scene can provide a rich source of evidence due to its increased storage capabilities and Internet connectivity. According to several news references, Ocsar Pistorius' iPads were examined by a mobile expert and presented during the murder trial to show Internet activity hours before the murder of his girlfriend. There are different ways to acquire forensic data from an iPhone. Though each method will have its positives and negatives, the fundamental principle of any acquisition method is to obtain a bit-by-bit picture of the original data.

This chapter covers physical acquisition techniques that target the physical storage medium directly and extract a disk image from the device into an external file, which can be examined later using forensic tools.

Operating modes of iOS devices

Before we dive into the forensic techniques and acquisition methods, it is important to know the different operating modes of an iPhone. Many forensic tools and methods require you to place the device into one of the operating modes. Understanding the iOS device operating modes is required in order to perform a particular action on the device. iOS devices are capable of running in different operating modes: normal mode, recovery mode, and DFU mode. Most forensic tools require the examiner to know which mode the device is currently utilizing. We will define each mode in this section. When the term "iPhone" is referenced, it should be understood that the statement remains true for all iOS devices.

Normal mode

When an iPhone is switched on, it is booted to its operating system. This mode is known as normal mode. Most of the regular activities (calling, texting, and so on) performed on an iPhone will be run in normal mode.

When an iPhone is turned on, internally, it goes through a **secure boot chain**, as shown in the following figure. Each step in the boot-up process contains software components that are cryptographically signed by Apple to ensure integrity.

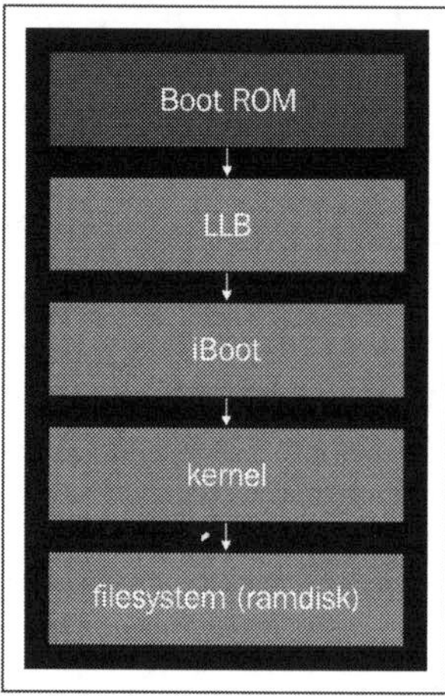

A secure boot chain of an iPhone in normal mode

The **Boot ROM**, known as the secure ROM, is a **read-only memory** (**ROM**) and is the first significant code that runs on an iPhone (`http://images.apple.com/ipad/business/docs/iOS_Security_Feb14.pdf`). The Boot ROM code contains the Apple root CA public key, which is used to verify the signature of the next stage before allowing it to load. When the iPhone is started, the application processor executes the code from the Boot ROM, which, in turn, verifies whether the **Low Level Bootloader** (**LLB**) is signed by Apple or not and loads it accordingly. When LLB finishes its tasks, it verifies and loads the second stage boot loader (iBoot). iBoot verifies and loads the iOS kernel, which, in turn, verifies and runs all the user applications as shown in the preceding figure. The secure boot chain ensures iOS runs only on validated Apple devices.

Recovery mode

During the boot-up process, if one step is unable to load or verify the next step, then the boot-up is stopped and the iPhone displays a screen, as shown in the following screenshot. This mode is known as the recovery mode. The recovery mode is required to perform upgrades or restore the iPhone.

To enter recovery mode, perform the following steps:

1. Turn off the device—press and hold down the **Sleep/Power** button located at the top of the iPhone until the red slider appears. Then, move the slider and wait for the device to turn off.
2. Hold down the iPhone **Home** button and connect the device to a computer via a USB cable. The device should turn on.
3. Continue holding the Home button until the **Connect to iTunes** screen appears, as shown in the following screenshot. Then, you can release the Home button. (On a jail-broken iOS device, this screen may appear with different icons.) Most forensic tools and extraction methods will alert the examiner to the current state of the iOS device.

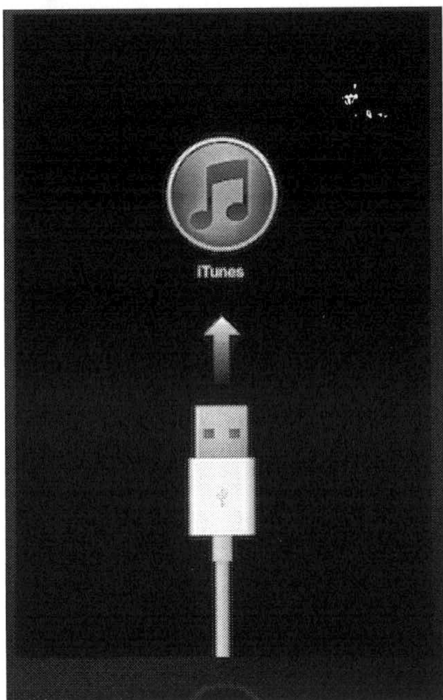

You can read about the iPhone recovery mode at `http://support.apple.com/kb/HT1808`.

To exit the recovery mode, reboot the iPhone. This can be completed by holding the Home and Sleep/Power button until the Apple logo appears. Normally, the process of rebooting returns the iPhone from recovery mode to normal mode. The examiner may experience a situation where the iPhone constantly reboots into recovery mode. This is known as a **recovery loop**. A recovery loop often occurs when the user attempts to jailbreak their iOS device and an error occurs.

Several open source methods exist to repair a recovery loop. The following example shows the **redsn0w** tool, which can be used to exit a recovery loop. You can download the latest version of redsn0w from the following link: `https://sites.google.com/a/iphone-dev.com/files/`.

Then, navigate to **Extras** | **Recovery fix,** as shown in the following screenshot. An external method or tool may not be required. Sometimes, placing the device in DFU mode and connecting the device to iTunes will properly reboot the iPhone.

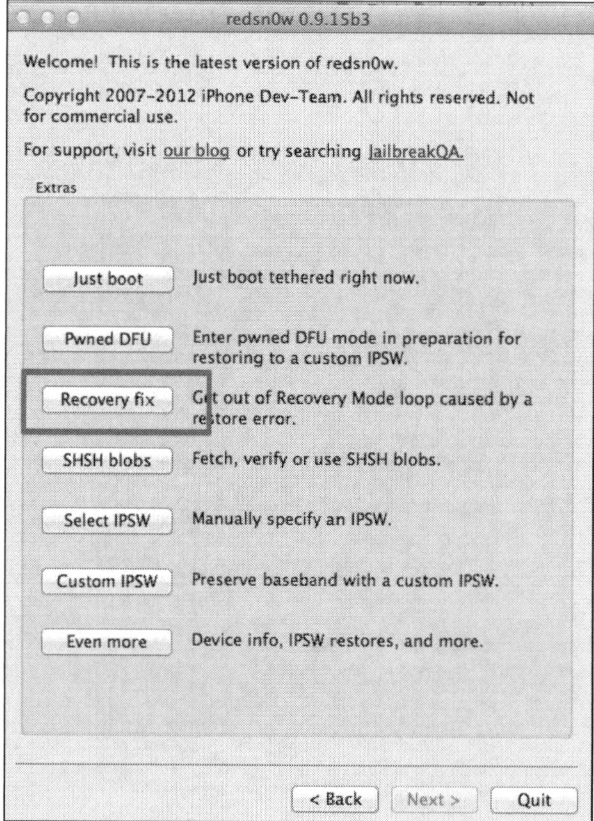

The redsn0w recovery fix

DFU mode

During the boot-up process, if the Boot ROM is not able to load or verify LLB, then the iPhone displays a black screen. This mode is known as the **Device Firmware Upgrade (DFU)** mode. DFU mode is a low-level diagnostic mode and is designed to perform firmware upgrades for the iPhone. During a firmware upgrade, the iPhone goes through a different boot sequence as shown in the following figure. Most forensic tools use DFU mode to perform a physical acquisition.

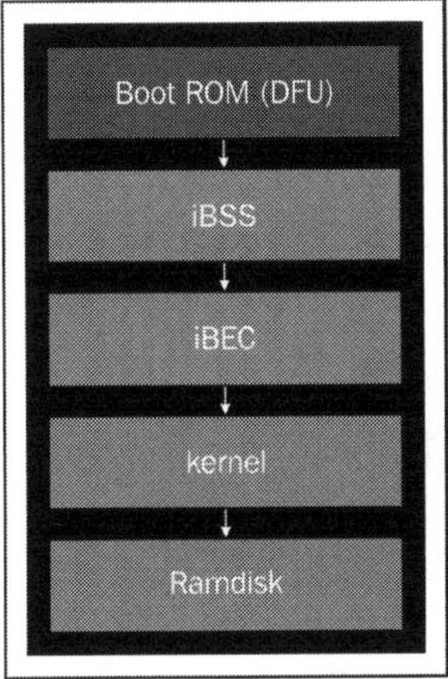

A secure boot chain of an iPhone in DFU mode

In DFU mode, the Boot ROM boots first, which, in turn, verifies and runs the second stage boot loaders, iBSS and iBEC. The iBEC loader verifies and loads the kernel. The kernel verifies and loads the **ramdisk** into memory. Again, most forensic acquisition methods require the iOS device to be successfully entered in DFU mode. As mentioned in *Chapter 1, Introduction to Mobile Forensics*, all steps must be well documented by the examiner. The handling of the iOS device is no exception. DFU mode is a method recognized in mobile device forensics and is deemed to be a forensically sound action to prepare the device for forensic acquisition.

To enter DFU mode, perform the following steps:

1. Download and install iTunes on your forensic workstation from `http://www.apple.com/itunes/download/`.
2. Connect your device to the forensic workstation via a USB cable.
3. Turn off the device.
4. Hold down the Power button for 3 seconds.
5. Hold down the Home button without releasing the Power button for exactly 10 seconds.

6. Release the Power button and continue to hold down the Home button until you are alerted by iTunes with the **iTunes has detected an iPhone in recovery mode. You must restore the iPhone before it can be used with iTunes** message.
7. At this point, the iPhone screen will be black and should not display anything. The iPhone is ready to be used in DFU mode. If you see the Apple logo or other signals that the device is booting, repeat steps 2 through 6 until iTunes displays that message.

To verify whether the iPhone is in DFU mode on Mac OS X, launch **System Information** and go to the **USB** option. You should see a device similar to what is shown in the following screenshot:

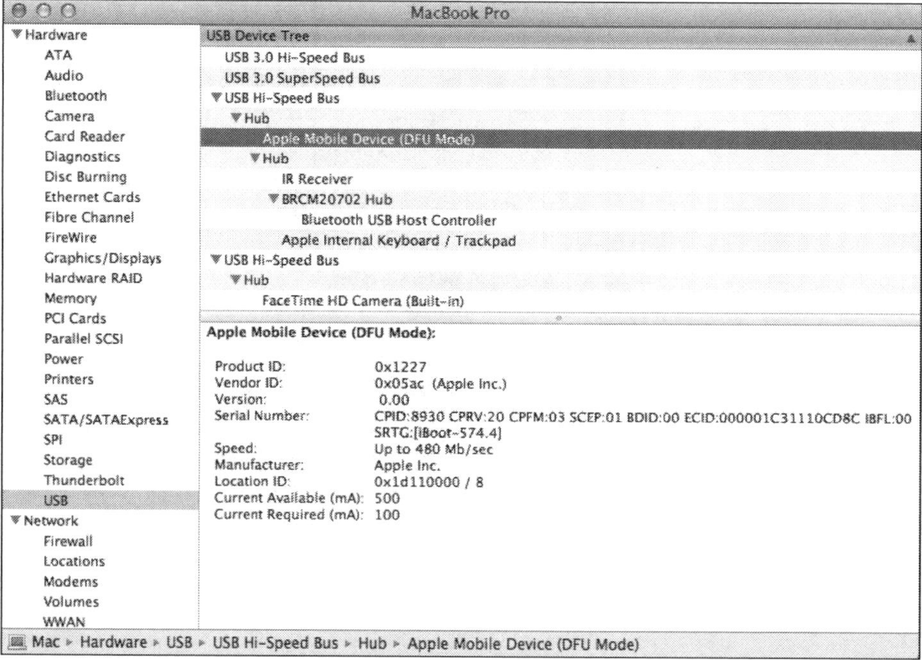

The MAC system information displaying a DFU-mode device

Just like in recovery mode, to exit DFU mode, hold down the Home button and the Power button until the Apple logo appears on the device. More information can be found on methods to verify DFU mode at `http://www.zdziarski.com/blog/wp-content/uploads/2013/05/iOS-Forensic-Investigative-Methods.pdf`.

Physical acquisition

iOS devices have two types of memory: volatile (RAM) and non-volatile (NAND Flash). RAM is used to load and execute the key parts of the operating system or the application. The data stored on the RAM is lost after a device reboots. RAM usually contains very important application information such as active applications, usernames, passwords, and encryption keys. Though the information stored in the RAM can be crucial in an investigation, currently there is no method or tool available to acquire the RAM memory from a live iPhone.

Unlike RAM, NAND is non-volatile memory and retains the data stored in it even after a device reboots. NAND flash is the main storage area and contains the system files and user data (`http://www.nist.gov/forensics/research/upload/draft-guidelines-on-mobile-device-forensics.pdf`). The goal of physical acquisition is to perform a bit-by-bit copy of the NAND memory, similar to the way in which a computer hard drive would be forensically acquired. While data storage seems similar, NAND differs from the magnetic media found in modern hard drives. NAND memory is cheaper, faster, and holds a great amount of data. Thus, NAND is the ideal storage for mobile devices as mentioned in *iPhone and iOS Forensics, Andrew Hoog And Katie Strzempka, Elsevier BV*.

Physical acquisition has the greatest potential for recovering data from iOS devices; however, evolving security features (secure boot chain, storage encryption, and passcode) on these devices may hinder the accessibility of the data during forensic acquisition. Researchers and commercial forensic tool vendors are continually attempting new techniques to bypass the security features and perform physical acquisition on iOS devices. Currently, there are two methods that can be used to gain access to the iOS device and grab a physical image of the NAND. The two methods are explained in detail in the following sections.

Acquisition via a custom ramdisk

Acquisition via a custom ramdisk is a novel method to acquire data from an iPhone. It gains access to the file system by loading a custom ramdisk into the memory and exploiting a weakness in the boot process while the device is in the DFU mode. A custom ramdisk contains the forensic tools necessary to dump the file system over USB via an SSH tunnel. Loading a custom ramdisk onto a device will not alter the user data, and thus the evidence will not be destroyed.

Imagine a computer that is protected with an OS-level password, we can still access the hard disk contents by booting with a live CD. Similarly, on the iPhone, we can load a custom ramdisk over USB and access the file system. However, the iPhone secure boot chain prevents us from loading the custom ramdisk. We can achieve this by exploiting a Boot ROM vulnerability and patching successive stages, as shown in the following figure:

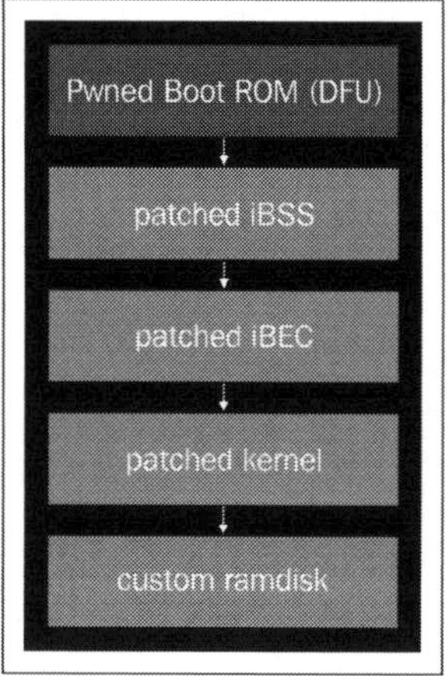

An exploited boot chain of an iPhone in DFU mode

Hacker communities have found several Boot ROM vulnerabilities in A4 devices (iPhone 4 and older iPhone models). Currently, there are no Boot ROM exploits for A5+ devices (iPhone 4S and later models) that allow access for physical acquisition of the device. Boot ROM vulnerabilities cannot be fixed with software updates, effectively making a device vulnerable forever.

In addition to this, the file system on the iPhone is encrypted. Since the release of the iPhone 3GS, the hardware and firmware encryption are built into iOS devices. Every iOS device has a dedicated AES 256-bit crypto engine (the AES cryptographic accelerator) with two hardcoded keys: UID (Unique ID) and GID (Group ID) (as stated by Zdziarski in one of his books). The CPU on the device cannot read the hardcoded keys but can use them for encryption and decryption through the AES accelerator. The UID key is unique for each device and is used to create device-specific keys (the `0x835` key and the `0x89B` key) that are later used for file system encryption. The UID allows data to be cryptographically tied to a particular device; so, even if the flash chip is moved from one device to other, the files are not readable and remain encrypted. The GID key is shared by all devices with the same application processor (for example, all devices that use the A4 chip) and is used to decrypt the iOS firmware images (IPSW) during installation, restore, and update. The GID prevents hackers from reversing the firmware and finding security vulnerabilities.

Apart from the UID and GID, all other cryptographic keys are created by the system's **random number generator** (**RNG**) using an algorithm based on Yarrow. More information on encryption and Yarrow-based algorithms can be found at `http://images.apple.com/ipad/business/docs/iOS_Security_Feb14.pdf`.

iPhone Data Protection Tools is an open source iOS forensic toolkit written by Jean-Baptiste and Jean Sigwald, which uses the custom ramdisk technique. The forensic toolkit builds a custom ramdisk and loads it to the device by exploiting the Boot ROM vulnerability in the DFU mode. The custom ramdisk includes tools to enumerate device information, brute force passcode attempts, and create a raw image of the disk partition. The forensic toolkit also obtains device encryption keys, decrypts the file system, and recovers the deleted files. The iPhone Data Protection Tools currently work with the iPhone 3G, 3GS and 4; iPod touch 2G, 3G and 4G; and iPad 1 models. More information on this can be found at `https://code.google.com/p/iphone-dataprotection/wiki/README`.

The forensic environment setup

The following steps explain how to use the iPhone Data Protection Tools on Mac OS X 10.8.5 with Xcode 4.6.1 and iOS 6.1 SDK (other versions should work with the same steps). Assuming that you already have Xcode with UNIX tools installed, you will need to install some additional command-line tools, Python modules, and binaries to build and use the iPhone Data Protection Tools.

Downloading and installing the ldid tool

First, you need to download the **ldid** tool, which is used to view and manipulate code signatures and embedded entitlements `plist` files of binaries. On Mac OS X, open the terminal window and use the `curl` command, as shown, to download the ldid tool:

```
$curl -O http://networkpx.googlecode.com/files/ldid
% Total % Received % Xferd Average Speed  Time  Time Time  Current
                                Dload  Upload Total Spent Left  Speed
100 32016 100 32016 0   0  52214     0 --:--:-- --:--:-- --:--:--  279k
```

Grant execution permission to the ldid tool and move it to the `bin` directory in the `usr` folder, using the commands shown:

```
$chmod +x ldid
$sudo mv ldid /usr/bin/
```

Verifying the codesign_allocate tool path

Create a symbolic link to the Xcode folder, as shown:

```
$sudo ln -s /Applications/Xcode.app/Contents/Developer /
```

iPhone Data Protection Tools require the `codesign_allocate` tool, which is present by default if the UNIX tools were installed with Xcode. To find whether `codesign_allocate` exists or not, use the command shown:

```
$which codesign_allocate/usr/bin/codesign_allocate
```

If you do not see the location of `codesign_allocate` from the command-line output, create a symbolic link to it, as shown:

```
$sudo ln -s/Developer/Platforms/iPhoneOS.platform/Developer/usr/bin/
codesign_allocate /usr/bin
```

Installing OSXFuse

iOS firmware files are in the `IMG3` file format. To modify the ramdisk, the iPhone Data Protection Tools include a FUSE file system that understands the `IMG3` format. The latest version of OSXFuse should be installed on your forensic workstation. OSXFuse extends the native file handling capabilities of OS X and allows you to mount the file systems that are not natively supported by OS X. You can download and install OSXFuse by executing the commands shown or directly from the following link: `http://sourceforge.net/projects/osxfuse/files/osxfuse-2.6.2/osxfuse-2.6.2.dmg`.

```
$sudo curl -O -L http://sourceforge.net/projects/osxfuse/files/
osxfuse-2.6.2/osxfuse-2.6.2.dmg
% Total % Received % Xferd Average Speed  Time  Time   Time  Current
                           Dload   Upload Total Spent  Left  Speed
100 8608k 100 8608k 0  0   546k      0  0:00:15 0:00:15 --:--:--698k
```

Next, run the three commands as shown:

```
$hdiutil mount osxfuse-2.6.2.dmg
Checksumming Gesamte Disk (Apple_HFS : 0)…
......................................................................
Gesamte Disk (Apple_HFS : 0): verified   CRC32 $6D4256E4
verified   CRC32 $D09075DF
/dev/disk2                                     /Volumes/FUSE for OS X
```

```
$sudo installer -pkg /Volumes/FUSE\ for\ OS\ X/Install\ OSXFUSE\ 2.6.pkg
-target /
installer: Package name is FUSE for OS X (OSXFUSE)
installer: Installing at base path /
installer: The install was successful.
$hdiutil eject /Volumes/FUSE\ for\ OS\ X/
"disk3" unmounted.
"disk3" ejected.
```

Installing Python modules

The Python scripts of iPhone Data Protection Tools require installation of several Python modules: `construct`, `progressbar`, and `setuptools`. You can install the required Python modules using Python's `easy_install` command, as shown:

```
$sudo easy_install construct progressbar
Searching for construct
Reading http://pypi.python.org/simple/construct/
Best match: construct 2.5.1
Downloading https://pypi.python.org/packages/source/c/construct/
construct-
2.5.1.zip#md5=4616eb3c12e86ba859ff2ed2f01ddb1c
Processing construct-2.5.1.zip
[...]
Installed /Library/Python/2.7/site-packages/construct-2.5.1-py2.7.egg
Processing dependencies for construct
Searching for six
Reading http://pypi.python.org/simple/six/
Best match: six 1.4.1
Downloading https://pypi.python.org/packages/source/s/six/six-1.4.1.tar.
gz#md5=bdbb9e12d3336c198695aa4cf3a61d62
Processing six-1.4.1.tar.gz
[...]
Installed /Library/Python/2.7/site-packages/six-1.4.1-py2.7.egg
Finished processing dependencies for construct
Searching for progressbar
Reading http://pypi.python.org/simple/progressbar/
Reading http://code.google.com/p/python-progressbar
Best match: progressbar 2.3
```

```
Downloading http://python-progressbar.googlecode.com/files/progressbar-
2.3.tar.gz
Processing progressbar-2.3.tar.gz
[...]
Installed /Library/Python/2.7/site-packages/progressbar-2.3-py2.7.egg
Processing dependencies for progressbar
Finished processing dependencies for progressbar
Searching for setuptools
Best match: setuptools 0.6c12dev-r88846
Adding setuptools 0.6c12dev-r88846 to easy-install.pth file
Installing easy_install script to /usr/local/bin
[...]
Processing dependencies for setuptools
Finished processing dependencies for setuptools
```

The Python scripts also require the cryptography modules **PyCrypto** and **M2Crypto** to decrypt iOS firmware images, files, and keychain items. You can download and install the PyCrypto tool directly from the following link: `https://rudix-mountainlion.googlecode.com/files/pycrypto-2.6-1.pkg`.

You can install the M2Crypto module using the commands shown:

```
$sudo curl -O -L http://chandlerproject.org/pub/Projects/MeTooCrypto/
M2Crypto-0.21.1-py2.7-macosx-10.8-intel.egg
% Total % Received % Xferd Average Speed Time  Time   Time  Current
                           Dload  Upload Total Spent  Left  Speed
100 477k 100 477k  0    0  63290    0  0:00:07 0:00:07 --:--:-- 102k
$sudo easy_install M2Crypto-0.21.1-py2.7-macosx-10.8-intel.egg
Processing M2Crypto-0.21.1-py2.7-macosx-10.8-intel.egg
[...]
Installed /Library/Python/2.7/site-packages/M2Crypto-0.21.1-py2.7-
macosx-10.8-
intel.egg
Processing dependencies for M2Crypto==0.21.1
Finished processing dependencies for M2Crypto==0.21.1
```

Finally, to download the latest copy of iPhone Data Protection Tools from the Google code repository, you need to install the **Mercurial** source code management system. You can download and install this using the `easy_install` command, as shown, or directly from the following link: `http://mercurial.berkwood.com/binaries/Mercurial-2.6.2-py2.7-macosx10.8.zip`.

```
$sudo easy_install mercurial
Searching for mercurial
Reading http://pypi.python.org/simple/mercurial/
Best match: mercurial 2.8
Downloading https://pypi.python.org/packages/source/M/Mercurial/
mercurial-
2.8.tar.gz#md5=76b565f48000e9f331356ab107a5bcbb
Processing mercurial-2.8.tar.gz
[...]
Processing dependencies for mercurial
Finished processing dependencies for mercurial
```

Downloading iPhone Data Protection Tools

Download the latest copy of iPhone Data Protection Tools using Mercurial (`hg`), as shown:

```
$sudo hg clone https://code.google.com/p/iphone-dataprotection/
warning: code.google.com certificate with fingerprint ad:3c:56:fb:
e8:c0:62:b0:ff:89:21:52:98:b1:a1:d4:94:a4:1c:84 not verified (check
hostfingerprints or web.cacerts config setting)
destination directory: iphone-dataprotection
requesting all changes
adding changesets
adding manifests
adding file changes
added 72 changesets with 2033 changes to 1865 files
updating to branch default
152 files updated, 0 files merged, 0 files removed, 0 files unresolved
```

The command in the preceding screenshot creates the `iphone-dataprotection` directory and downloads iPhone Data Protection Tools to it.

Building the IMG3FS tool

Build the IMG3 FUSE file system from the `img3fs` directory. This module enables you to directly mount the firmware disk images included in the iOS firmware packages (IPSW), as shown in the following command lines:

```
$cd iphone-dataprotection

$sudo make -C img3fs/
gcc -o img3fs img3fs.c -Wall -lfuse_ino64 -lcrypto -
I/usr/local/include/osxfuse || gcc -o img3fs img3fs.c -Wall -
losxfuse_i64 -lcrypto -I/usr/local/include/osxfuse

img3fs.c: In function 'img3_check_decrypted_data':
img3fs.c:100: warning: pointer targets in passing argument 2 of
'strncmp' differ in signedness
img3fs.c:104: warning: pointer targets in passing argument 2 of
'strncmp' differ in signedness
img3fs.c:108: warning: pointer targets in passing argument 2 of
'strncmp' differ in signedness
[...]
```

After running the `make` command, you will notice a few compiler warning messages, which you can ignore.

Downloading redsn0w

Firmware disk images included in the iOS firmware packages are encrypted. The **redsn0w** application, a famous iOS jailbreaking utility developed by the iPhone Dev Team, contains a `plist` file with the decryption keys for all previously released iOS firmware images. The iPhone Data Protection build scripts will use the decryption keys to automatically decrypt the kernel and ramdisk. To do this, download the latest version of redsn0w and create a symbolic link to its `Keys.plist` file in the current directory, as shown in the following code. Later in this chapter, you will also use redsn0w to boot the custom ramdisk onto the device.

```
$sudo curl -O -L https://sites.google.com/a/iphone-
dev.com/files/home/redsn0w_mac_0.9.15b3.zip

% Total % Received % Xferd Average Speed   Time Time  Time  Current
Dload   Upload Total Spent Left  Speed

100 17.1M 100 17.1M 0    0    298k      0  0:00:58 0:00:58 --    329k

$sudo unzip redsn0w_mac_0.9.15b3.zip

Archive:  redsn0w_mac_0.9.15b3.zip

creating: redsn0w_mac_0.9.15b3/

inflating: redsn0w_mac_0.9.15b3/boot-ipt4g.command

inflating: redsn0w_mac_0.9.15b3/credits.txt
```

```
inflating: redsn0w_mac_0.9.15b3/license.txt
[...]
extracting: redsn0w_mac_0.9.15b3/redsn0w.app/Contents/PkgInfo
creating: redsn0w_mac_0.9.15b3/redsn0w.app/Contents/Resources/
inflating:
redsn0w_mac_0.9.15b3/redsn0w.app/Contents/Resources/redsn0w.icns
$sudo cp redsn0w_mac_0.9.15b3/redsn0w.app/Contents/MacOS/Keys.plist .
```

Creating and loading the forensic toolkit

At this point, all of the prerequisites should be installed, and you should be ready to build and load the custom ramdisk onto your target iOS device. First, we patch the ramdisk signature checks in the kernel and build a custom ramdisk with our forensic tools. Later, we use redsn0w to load the modified kernel and the custom ramdisk by exploiting the Boot ROM vulnerability.

Downloading the iOS firmware file

An iOS firmware update software archive (IPSW) file for the hardware model with which you intend to use the custom ramdisk is required. iPhone Data Protection Tools supports the ramdisk creation for iOS 6 IPSW and lower versions. For best results, use the latest version of iOS 5 IPSW to create the ramdisk. iOS 5 kernel is compatible with the previous and forthcoming iOS versions. So, even if your device is running on iOS 7 or iOS 4, you can prepare the ramdisk with iOS 5. You can download the IPSW file for the target device from `http://getios.com/index.php`.

Copy the downloaded IPSW to the `dataprotection` directory inside the `iphone` folder, as shown in the following command:

```
$cp ~/Downloads/iPhone3,1_5.1.1_9B208_Restore.ipsw .
```

The above command ends with `.` which represents the current working directory.

The `iPhone3,1_5.1.1_9B208_Restore.ipsw` file used in the preceding command targets the iPhone 4 device. The IPSW filenames include the hardware model (iPhone3,1), the iOS version number (5.1.1), and the specific build number (9B208).

Modifying the kernel

For the custom ramdisk to work properly, a modified kernel is required. The `kernel_patcher.py` script in iPhone Data Protection Tools extracts the `kernelcache` from the supplied IPSW file and patches it. The kernel patching utility makes appropriate changes to the kernel to disable the code signing to run arbitrary binaries and to allow access to restricted functions. Run the `kernel_patcher.py` script on your IPSW to create a patched `kernelcache` and a shell script that builds the ramdisk, as shown in the following commands:

```
$sudo python python_scripts/kernel_patcher.py iPhone3,1_5.1.1_9B208_
Restore.ipsw
Decrypting kernelcache.release.n90
Unpacking ...
Doing CSED patch
Doing getxattr system patch
Doing nand-disable-driver patch
Doing task_for_pid_0 patch
Doing IOAES gid patch
Doing AMFI patch
Doing _PE_i_can_has_debugger patch
Doing IOAESAccelerator enable UID patch
Patched kernel written to kernelcache.release.n90.patched
Created script make_ramdisk_n90ap.sh, you can use it to (re)build the
ramdisk
```

The script creates a patched kernel file called `kernelcache.release.n90.patched` to the current working directory. For the iOS 5 IPSW file, it also creates a script called `make_ramdisk_n90ap.sh` to build the custom ramdisk. Pay attention to the file names because they may differ depending on the iOS device model.

Building a custom ramdisk

Give permission to execute the `make_ramdisk_n90ap.sh` ramdisk build script and execute this script to create the custom ramdisk as follows:

```
$chmod +x make_ramdisk_n90ap.sh
```

Before executing the script, edit the file and fix the iOS SDK path as follows:

```
$sudo nano make_ramdisk_n90ap.sh
```

As we are using iOS SDK 6.1, append 6.1 to the `for` loop, as shown in the following code:

```
for VER in 4.2 4.3 5.0 5.1 6.0 6.1
```

Fix the IOKit path by replacing `/Developer/Platforms/iPhoneOS.platform/Developer/SDKs/iPhoneOS$VER.sdk/System/Library/Frameworks/IOKit.framework/IOKit` with `/Applications/Xcode.app/Contents/Developer/Platforms/iPhoneOS.platform/Developer/SDKs/iPhoneOS$VER.sdk/System/Library/Frameworks/IOKit.framework/IOKit` in the `if` statement.

After making the necessary changes, press *Ctrl* + *X*, type the letter `y` and hit the *Enter* key on the keyboard to save the file.

Execute the `make_ramdisk_n90ap.sh` script; it will download `ssh.tar.gz` from Google Code. Next, compile the ramdisk tools located in the `ramdisk_tools` folder and add them to the existing ramdisk to prepare a forensic ramdisk, as shown in the following command:

```
$sudo ./make_ramdisk_n90ap.sh
Found iOS SDK 6.1
[some warning messages]
Archive:  iPhone3,1_5.1.1_9B208_Restore.ipsw
inflating: 038-5512-003.dmg
TAG: TYPE OFFSET 14 data_length:4
[...]
"disk2" unmounted.
"disk2" ejected.
You can boot the ramdisk using the following command (fix paths)
redsn0w -i iPhone3,1_5.1.1_9B208_Restore.ipsw -r myramdisk_n90ap.dmg
-k kernelcache.release.n90.patched
Add -a "-v rd=md0 nand-disable=1" for nand dump/read only access
```

If you are using an iOS 6 IPSW file, run the `build_ramdisk_ios6.sh` file to create the custom ramdisk. Before running the script, you need to edit `Makefile` in the `ramdisk_tools` directory, fix the iOS SDK version, and compile it using the `make` command.

Booting the custom ramdisk

To load the custom ramdisk onto the device, start redsn0w from the command line using the IPSW, custom ramdisk, and patched kernel as shown in the following command:

```
$sudo ./redsn0w_mac_0.9.15b3/redsn0w.app/Contents/MacOS/redsn0w -i
iPhone3,1_5.1.1_9B208_Restore.ipsw -r myramdisk_n90ap.dmg -k
kernelcache.release.n90.patched
```

Turn off your iOS device and connect it to the computer, which is running redsn0w, with a USB cable. When the device is connected, redsn0w is displayed on the screen as shown in the following screenshot:

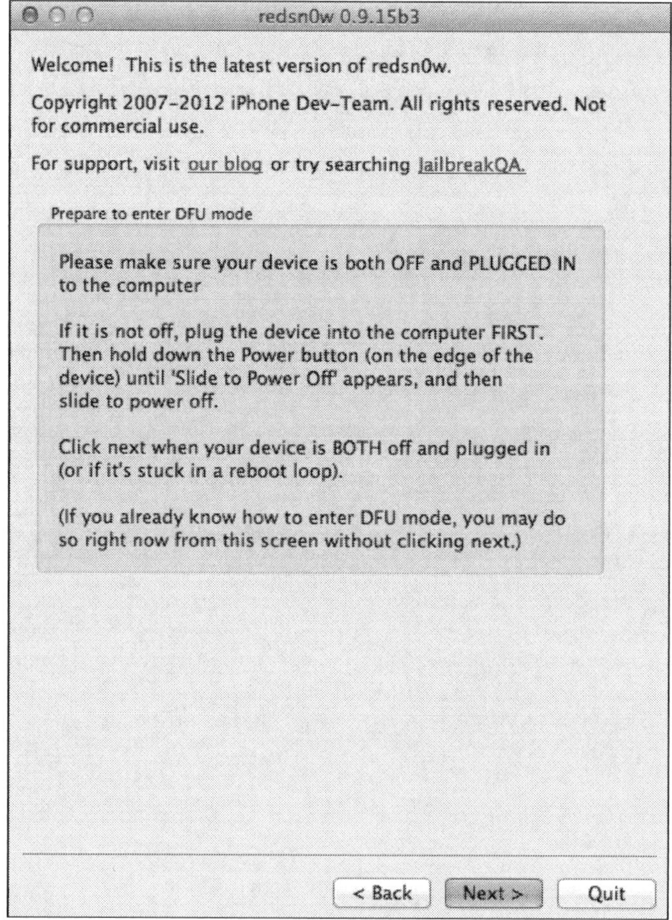

The redsn0w welcome screen

Click on **Next** and follow the steps displayed on the screen to place the device in the DFU mode. Once your device is in the DFU mode, redsn0w exploits one of the Boot ROM vulnerabilities and loads the modified kernel and custom ramdisk. If the process is successful, you will notice the image of a pineapple on the iPhone, followed by boot messages in small text. Once the process is completed, you will notice an ASCII version of the **OK** message on the device.

Establishing communication with the device

The custom ramdisk booted onto the iPhone contains an SSH server, which will allow remote command-line access to the device through the USB protocol. The USB multiplexing daemon (`usbmuxd`), a background daemon in Apple's mobile device framework, is used to tunnel the TCP socket connection over the USB protocol to a local TCP socket listening on the device. In this case, run `tcprelay.py`, as shown in the following command line, to connect to the SSH server that is running on the custom ramdisk:

```
$python usbmuxd-python-client/tcprelay.py -t 22:2222 1999:1999
Forwarding local port 2222 to remote port 22
Forwarding local port 1999 to remote port 1999
```

Other python scripts included in iPhone Data Protection Tools communicate with the device over SSH. So, you should keep running `tcprelay.py` in another terminal until you acquire data from the device.

Bypassing the passcode

The iPhone provides an option for its users to set a passcode on their device to prevent unauthorized access. Once a passcode is set, whenever the device is turned on or awakened from sleep mode, the passcode is required to access the data. iOS supports a simple four-digit code and complex alphanumeric passcodes of any length. With the iPhone 5S, the user fingerprint scan can also be used to lock/unlock the device. For iPhone 5S, the user can also select a simple four-digit code to use in case the fingerprint is not recognized. By default, the passcode is a four-digit numeric code but by modifying the settings, it can be set to be a complex passcode. The user also has the option to erase all the contents on the iPhone after 10 failed passcode attempts.

Passcode-locked devices are being utilized more frequently due to general user awareness of theft and security policies from organizations. Circumventing the passcode is not always possible due to security improvements in iOS. The forensic examiner should try to secure the passcode from the owner to prevent issues in acquiring data from newer, locked iOS devices.

In the initial releases of iOS until iOS 3, the passcode for unlocking the device was stored directly in the keychain, a place to store passwords securely on the iPhone. This passcode security can be bypassed by just removing the record from the keychain or by removing the UI setting that asks for the passcode after booting with the custom ramdisk.

Since iOS 4, the passcode is not stored on the device in any format. By setting a device passcode, the user automatically enables data protection, which protects the data at rest. With data protection, the data on the device is encrypted with a set of class keys stored in the **System keybag**. The System keybag itself is protected with a passcode key, generated from the user's passcode and the device's UID. So, in order to decrypt the protected keychain items and files on the file system, you first need to decrypt the System keybag. If there is no passcode, the System keybag can be easily decrypted. If there is a simple four-digit passcode, you will have to guess it to decrypt the System keybag. As the passcode is tangled with the device's UID key, brute force attempts must be performed on the device. Also, the same passcode on different devices generates different passcode keys as the UID is unique per device. Passcode brute force attacks performed at the springboard level introduce delays, lock the device, and may lead to the wiping of data. However, these protection mechanisms are not applicable when you are performing a brute force attack on a kernel extension (AppleKeyStore) to decrypt the System keybag. It is worth mentioning that some tools will attempt to crack the passcode on an iOS device by accessing the host computer for which that iOS device was connected and synced. The tool accesses the pairing key through an escrow file to decrypt the locked device. For this to work, the examiner would need to have access to both the iOS device and the host computer to which the device is backed up.

Should the host computer not be available, as mentioned, the `demo_bruteforce.py` Python script included in iPhone Data Protection Tools can perform brute force attack and guess any four-digit passcode within 18 minutes. Brute force on the device is slow, and the time required to brute force a passcode depends on the device's capability. The following table lists the time required to brute force passcodes of various lengths and complexity requirements on the iPhone 4:

Passcode length	Complexity	Time
4	Numeric	18 minutes
4	Alphanumeric	19 days
6	Alphanumeric	196 days
8	Alphanumeric	755 thousand years
8	Alphanumeric, complex	27 million years

On Mac OS X, open a new terminal and run the following command. The brute force script uses the `1999` port opened with `tcprelay.py` to communicate with the ramdisk tools on the device. The script brute forces the passcode, decrypts the System keybag, dumps the data protection keys, and places them into a directory named with the **Unique Device Identifier** (**UDID**) of the target device in a `.plist` format.

```
$sudo python python_scripts/demo_bruteforce.py
Connecting to device : b716de79051ef093a98fc3ff1c46ca5e36faabc3
Keybag UUID : 5b14620bd1e74013bfa66325b6946773
Enter passcode or leave blank for bruteforce:
```

Hit *Enter* on the keyboard to start the brute force process:

```
Trying all 4-digits passcodes...
0 of 10000 ETA:  --:--:--
10 of 10000 ETA:  0:30:48
20 of 10000 ETA:  0:30:33
30 of 10000 ETA:  0:30:18
40 of 10000 ETA:  0:30:02
50 of 10000 ETA:  0:29:51
1100 of 10000 ETA:  0:25:54
1110 of 10000 ETA:  0:25:53
10000 of 10000 Time: 0:03:14
100% |############################################################|
BruteforceSystemKeyBag : 0:03:14.543986
{'passcode': '1111', 'passcodeKey':
'1f5c25823297f97f3cb38d998726fc22787ca3f31b8932c2b868700a341145b5'}
True
Keybag type : System keybag (0)
Keybag version : 3
Keybag UUID : 5b14620bd1e74013bfa66325b6946773
-------------------------------------------------------------------
Class        WRAP                 Type        Key        Public key
-------------------------------------------------------------------
NSFileProtectionComplete                                  3     AES
746f01658ec28b3ba99339e35beb37232f89658fd0214eb4c4cac99976b05039
NSFileProtectionCompleteUnlessOpen                        3     Curve25519
65db69526ea4026227d5faa0dc9066c1092e510aa586a2f62d9101e419600703
a035e0f5a6ee59b9e5928cc67b644c6a5cc8c5235c1a5440a02686d222fc3a08
NSFileProtectionCompleteUntilFirstUserAuthentication 3    AES
a32826f0abdf6fb1c049d395baa12b07e05a310fb49626a5cef078ca4a7a46f4
NSFileProtectionRecovery?                                 3     AES
28ec11f7719c7b36d6f4621a07c3b088fe65c9909c7adb45cf73ad8b9814a330
```

```
kSecAttrAccessibleWhenUnlocked                        3     AES
bab62b621ebcf0fbc97ee9a2f1fb6d3ee4a198f5a49a7e233c9dcdf2805292e0
kSecAttrAccessibleAfterFirstUnlock                    3     AES
638ae8c4a1a694b8db2968eba28ef39a14d5397ef102e4872395df619bd00d31
kSecAttrAccessibleAlways                              1     AES
5071e2058e148b7deee5b08fd685c0b29cd9d717f57732647dee0239513c7c79
kSecAttrAccessibleWhenUnlockedThisDeviceOnly          3     AES
3702f4d05b3b910860b9f17577d5f34bbf26e9a6f20594ea308d72919e182531
kSecAttrAccessibleAfterFirstUnlockThisDeviceOnly      3     AES
3d8fbd6b41c520f1dc8ebe6786abe4848fa1799456300b89c630c23ff931d6c8
kSecAttrAccessibleAlwaysThisDeviceOnly                1     AES
1774408c99198fb048ca5fbcd06feadc7d5e4c28a571111df557db9f58040ba5
[...]
```

If the user chooses a strong passcode that is not easy to guess, we can still access the files protected with `NSFileProtectionNone` and keychain items protected by the `kSecAttrAccessibleAlways` data protection classes.

Imaging the data partition

Physical imaging refers to the `dd` image of the logical partitions. As discussed in *Chapter 2, Understanding the Internals of iOS Devices*, NAND flash on iOS devices contains two logical disk partitions: system partition and user data partition. On a non-jailbroken device, the system partition will be kept in the read-only format. The user data partition contains all the user-installed applications and data. For full forensic analysis, it is preferred that both the system and data partition are acquired. Most forensic tools will capture both partitions in one image. If the examiner has a time crunch, at the minimum, they should dump the entire data partition. To acquire a disk image of the user data partition, run the `dump_data_partition.sh` shell script, as shown in the following command lines:

```
$sudo ./dump_data_partition.sh
Warning: Permanently added '[localhost]:2222' (RSA) to the list of
known hosts.
root@localhost's password:
```

Enter `alpine` as the password, which is the default SSH password on iOS devices, and hit *Enter* on the keyboard:

```
Device UDID : b716de79051ef093a98fc3ff1c46ca5e36faabc3
Dumping data partition in
b716de79051ef093a98fc3ff1c46ca5e36faabc3/data_20131209-1956.dmg ...
```

```
Warning: Permanently added '[localhost]:2222' (RSA) to the list of
known hosts.
[...]
```

The raw disk image will begin transferring, as shown in the following command lines, which should also be reflected by a gradual increase in the size of the file on the desktop. The script runs for several minutes to hours depending on the size of the file system. For example, acquiring an image from an 8 GB iPhone 4 roughly takes 30 minutes.

```
1801554+0 records in
1801554+0 records out
14758330368 bytes (15 GB) copied, 2463.01 s, 6.0 MB/s
```

The script dumps the entire user data partition and places it into a directory named `UDID` of the target device in a `DMG` format that can be mounted directly onto Mac OS X. Only the user data partition is copied, so the actual file size will be less than the iPhone size. Double-clicking on the `DMG` file mounts it in read-write mode and might effect the image integrity. To maintain the integrity, you can use the `hdiutil` command to mount the image in read-only mode, as shown in the following command. (Note that the file path reflects the `DMG` file you created.)

```
$hdiutil attach -readonly
b716de79051ef093a98fc3ff1c46ca5e36faabc3/data_20131209-1956.dmg
/dev/disk3                                              /Volumes/Data
```

The output of the `hdiutil` command shows that the disk image has been attached to the `/dev/disk3` device file and can be mounted on `/Volumes/Data` with the following command:

```
$cd /Volumes/Data/
```

You can now browse the file system in `/Volumes/Data/` and observe that all file contents are encrypted, as shown in the following command:

```
$hexdump -C mobile/Library/AddressBook/AddressBook.sqlitedb | head
```

The output is as shown in the following screenshot:

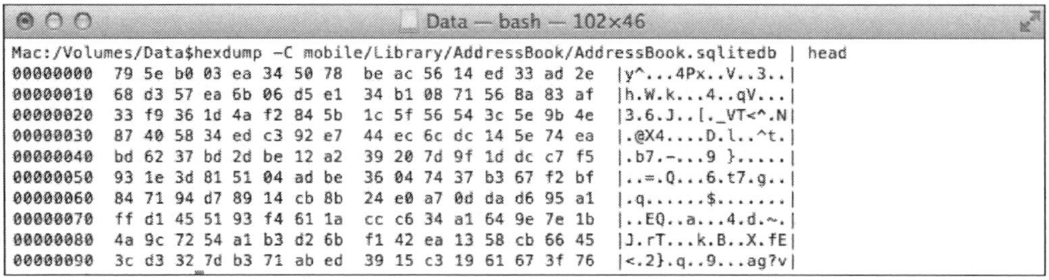

```
Data — bash — 102×46
Mac:/Volumes/Data$hexdump -C mobile/Library/AddressBook/AddressBook.sqlitedb | head
00000000  79 5e b0 03 ea 34 50 78  be ac 56 14 ed 33 ad 2e  |y^...4Px..V..3..|
00000010  68 d3 57 ea 6b 06 d5 e1  34 b1 08 71 56 8a 83 af  |h.W.k...4..qV...|
00000020  33 f9 36 1d 4a f2 84 5b  1c 5f 56 54 3c 5e 9b 4e  |3.6.J..[._VT<^.N|
00000030  87 40 58 34 ed c3 92 e7  44 ec 6c dc 14 5e 74 ea  |.@X4....D.l..^t.|
00000040  bd 62 37 bd 2d be 12 a2  39 20 7d 9f 1d dc c7 f5  |.b7.-...9 }.....|
00000050  93 1e 3d 81 51 04 ad be  36 04 74 37 b3 67 f2 bf  |..=.Q...6.t7.g..|
00000060  84 71 94 d7 89 14 cb 8b  24 e0 a7 0d da d6 95 a1  |.q......$.......|
00000070  ff d1 45 51 93 f4 61 1a  cc c6 34 a1 64 9e 7e 1b  |..EQ..a...4.d.~.|
00000080  4a 9c 72 54 a1 b3 d2 6b  f1 42 ea 13 58 cb 66 45  |J.rT...k.B..X.fE|
00000090  3c d3 32 7d b3 71 ab ed  39 15 c3 19 61 67 3f 76  |<.2}.q..9...ag?v|
```

The encrypted addressBook file

To unmount the image, use the `hdiutil eject` command as follows:

```
$cd /
$hdiutil eject /Volumes/Data/
"disk3" unmounted.
"disk3" ejected.
```

When the extracted disk image is mounted on Mac OS X, you can browse the file system. However, you cannot read the files as they are encrypted. To read any file data, the file contents must be decrypted using the keys in the System keybag.

Decrypting the data partition

The entire file system is encrypted with an EMF key, with the exception of actual files on the file system, which are encrypted with other keys (the data protection class keys). The EMF key is encrypted with the `0x89B` key. The `emf_decrypter.py` Python script included in iPhone Data Protection Tools can be used to decrypt the raw disk image. This script uses the raw disk image and keys in the aforementioned plist to decrypt all of the encrypted files on the file system, as shown in the following command lines:

```
$sudo python python_scripts/emf_decrypter.py
b716de79051ef093a98fc3ff1c46ca5e36faabc3/data_20131209-1956.dmg
b716de79051ef093a98fc3ff1c46ca5e36faabc3/f03d282cc7182d46.plist
Password:
Using plist file
b716de79051ef093a98fc3ff1c46ca5e36faabc3/f03d282cc7182d46.plist
Keybag unlocked with passcode key
```

```
cprotect version : 4 (iOS 5)
Test mode : the input file will not be modified
Press a key to continue or CTRL-C to abort
```

Hit *Enter* to continue the script execution:

```
Decrypting iNode1559014
Decrypting iNode3056993
Decrypting iNode3056996
Decrypting iNode6811
[...]
Decrypting AddressBook.sqlitedb
Decrypting AddressBook.sqlitedb-shm
Decrypting AddressBook.sqlitedb-wal
Decrypting AddressBookImages.sqlitedb
Decrypting AddressBookImages.sqlitedb-shm
[...]
Decrypting IMG_1117.JPG
Decrypting IMG_1128.PNG
Decrypting IMG_1139.JPG
[...]
Decrypting KeywordIndex.plist
Decrypting Manifest.sqlitedb
Decrypting express.psa
Decrypted 50518 files
```

The script modifies the disk image directly and the files are now decrypted and readable. To verify this, you can mount the disk image and examine `AddressBook.sqlitedb`, which was previously unreadable, with the following command:

```
$hdiutil attach -readonly data_20131209-1956.dmg
/dev/disk3                                    /Volumes/Data
$cd /Volumes/Data/
$hexdump -C mobile/Library/AddressBook/AddressBook.sqlitedb | head
```

The output is as shown in the following screenshot:

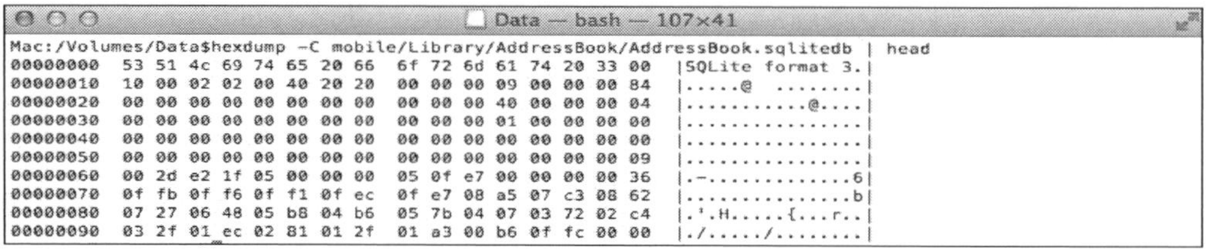

```
Data — bash — 107×41
Mac:/Volumes/Data$hexdump -C mobile/Library/AddressBook/AddressBook.sqlitedb | head
00000000  53 51 4c 69 74 65 20 66  6f 72 6d 61 74 20 33 00  |SQLite format 3.|
00000010  10 00 02 02 00 40 20 20  00 00 00 09 00 00 00 84  |.....@  ........|
00000020  00 00 00 00 00 00 00 00  00 00 00 40 00 00 00 04  |...........@....|
00000030  00 00 00 00 00 00 00 00  00 00 00 01 00 00 00 00  |................|
00000040  00 00 00 00 00 00 00 00  00 00 00 00 00 00 00 00  |................|
00000050  00 00 00 00 00 00 00 00  00 00 00 00 00 00 00 09  |................|
00000060  00 2d e2 1f 05 00 00 00  05 0f e7 00 00 00 00 36  |.-.............6|
00000070  0f fb 0f f6 0f f1 0f ec  0f e7 08 a5 07 c3 08 62  |...............b|
00000080  07 27 06 48 05 b8 04 b6  05 7b 04 07 03 72 02 c4  |.'.H.....{...r..|
00000090  03 2f 01 ec 02 81 01 2f  01 a3 00 b6 0f fc 00 00  |./...../........|
```

The decrypted AddressBook file

Now, you should be able to fully examine the artifacts on the data partition, which will be covered in detail in *Chapter 5, iOS Data Analysis and Recovery*.

Recovering the deleted data

Once a raw image of the device is obtained, you can recover the deleted files in the unallocated space by carving the HFS journal using the `emf_undelete.py` script. This script recovers only a limited number of files, as shown in the following command:

```
$sudo python python_scripts/emf_undelete.py UDID/data_20131209-
1956.dmg
```

To recover more deleted files or photos, acquire a low-level NAND image using `ios_examiner.py` and run the `undelete` command.

To acquire a low-level NAND image, boot the custom ramdisk and the patched kernel onto the iPhone with the `nand-disable` boot flag, as shown in the following command:

```
$sudo ./redsn0w_mac_0.9.15b3/redsn0w.app/Contents/MacOS/redsn0w -i
iPhone3,1_5.1.1_9B208_Restore.ipsw -r myramdisk_n90ap.dmg -k
kernelcache.release.n90.patched -a "-v rd=md0 nand-disable=1"
```

Once the ramdisk is booted successfully, run the `ios_examiner.py` script without parameters. It allows you to enter commands in the `ios_examiner` shell, as shown in the following command lines:

```
$cd iphone-dataprotection
$sudo python python_scripts/ios_examiner.py
Connecting to device : b716de79051ef093a98fc3ff1c46ca5e36faabc3
Device model: iPhone 4 GSM
UDID: b716de79051ef093a98fc3ff1c46ca5e36faabc3
ECID: 1937316564364
```

```
Serial number: 870522V6A4S
key835: ef8f36fb3a85b42a72e8c5efa6b1a844
key89B: de75b5f5fa6abc5bf25293b38f980a52
[...]
YaFTL_readCxtInfo FAIL, restore needed maxUsn=4491408
FTL restore in progress
100% |########################################|
BTOC not found for block 13 (usn 4491530), scanning all pages
402 used pages in block
LwVM header CRC OK
cprotect version : 4 (iOS 5)
iOS version:  5.1.1
Keybag state: locked
(iPhone4-data) /
```

Run the `bruteforce` command to brute force the passcode and unlock the keybag:

```
(iPhone4-data) / bruteforce
Passcode comlexity (from OpaqueStuff) : 4 digits
Enter passcode or leave blank for bruteforce:
```

Hit *Enter* and you will see the following command lines:

```
Passcode "" OK
Keybag state: unlocked
Save device information plist to [b716de7905.plist]:
```

Hit *Enter* to save the encryption keys to a plist file (`b716de7905.plist`).

Run the `nand_dump` command as shown in the following command lines. It copies the NAND image to the `dataprotection` folder.

```
(iPhone4-data) / nand_dump iphone4-nand.bin
Dumping 16GB NAND to iphone4-nand.bin
100% |############################################|
NAND dump time : 0:45:36.200233
SHA1: a16aa578679ef6a787c8c26a40de4b745a3ae179
```

Once the NAND image and the plist file are obtained, you can use `ios_examiner.py` and run the `undelete` command to recover the deleted files, as shown in the following command lines:

```
$sudo python python_scripts/ios_examiner.py iphone4-nand.bin b716de7905.
plist
Loading device information from b716de7905.plist
Device model: iPhone 4 GSM
UDID: b716de79051ef093a98fc3ff1c46ca5e36faabc3
ECID: 1937316564364
Serial number: 870522V6A4S
key835: ef8f36fb3a85b42a72e8c5efa6b1a844
key89B: de75b5f5fa6abc5bf25293b38f980a52
[...]
cprotect version : 4 (iOS 5)
iOS version:  5.1.1
(iPhone4-data) / undelete
Building FTL lookup table v1
100% |####################################|
Collecting existing file ids
23297 file IDs
Carving catalog file
Found deleted file record 51657 shaders.data created 2012-06-09
02:19:28
Found deleted file record 51656 shaders.maps created 2012-06-09
02:19:28
[...]
Carving attribute file for file keys
20261 files, 50997 keys
_FBStory.h
[...]
```

The command recovers the deleted files and places them into a directory named `undelete`. The recovery process is slow and takes hours to recover all the files.

If a device is restored, wiped, or upgraded to a new OS version, the file system key (EMF) is erased and a new key is recreated. Without the original EMF key, it is not possible to recover the underlying file system structure. So, it is not possible to recover the deleted files when an iPhone is restored, wiped, or upgraded. Also, iOS devices include a feature called **Effaceable Storage** to securely erase the keys. This feature accesses the underlying storage (NAND) to directly address and erase a small number of blocks at a very low level, which makes it impossible to recover deleted keys.

Acquisition via jailbreaking

To perform physical acquisition on devices that are not vulnerable to the Boot ROM exploit, the device must be jailbroken. Jailbreaking an iPhone allows the examiner to install tools that would not normally be on the device, such as SSH. By far, the most popular method for jailbreaking is with redSn0w or evasi0n. Both tools have simple wizards that will step the iOS device through the jailbreak process and install the **Cydia** application. An examiner should only jailbreak a device as a last resort and should use great caution when doing so. Again, all steps taken by the examiner must be well-documented. The jailbreaking process makes changes to the device, which may damage evidence or render it inadmissible in court. If possible, consider performing a **logical acquisition** first to preserve evidence that may be lost during the jailbreaking process.

To obtain an image of the user data partition, the forensic workstation and the target iOS device must be placed on the same wireless network. From the forensic workstation, run the following SSH command to start the process. Make sure that you replace the IP address used in the command with your device's IP address before running it.

```
$ssh root@192.168.2.9 "dd if=/dev/rdisk0s1s2  bs=8192" > data.dmg
```

Enter `alpine` as the password and hit *Enter* on the keyboard. This process may take several hours depending on the capacity of the iPhone. Once completed, it displays a certain number of bytes that have been copied, as shown in the following command lines:

```
1801554+0 records in
1801554+0 records out
14758330368 bytes (15 GB) copied, 2722.38 s, 5.4 MB/s
```

The SSH command connects to the SSH server on the iOS device as a root user. The `dd if=/dev/rdisk0s1s2 bs=8192` command executes the disk copy utility on the iPhone and reads the user data partition located at `/dev/rdisk0s1s2` with a block size of 8K. The command outputs the `data.dmg` file onto the forensic workstation drive. The resulted image file can be manipulated by the forensic analyst's choice of tools.

It is not possible to jailbreak a device that is protected with a passcode. So, if a device (A5+) is protected with a passcode and is not jailbroken, it is not possible to perform physical acquisition on that device. Also, it should be noted that the raw disk image obtained from the iPhone is encrypted and cannot be parsed. In order to decrypt the image, we must obtain encryption keys from the device. The encryption keys are tied to the device's UID key, which can be used only when the IOAESAccelerator kernel extension is patched. It is easy to obtain encryption keys on devices that run on iOS 5 and earlier versions. Since iOS 6, Apple introduced new security features to the kernel such as **Kernel Address Space Layout Randomization** and **Kernel Address Space Protection**, which prevent examiners from patching the kernel code directly. However, the Elcomsoft iOS Forensic Toolkit, a commercial tool for iOS forensics, claims that it is capable of performing physical acquisition on devices that run on iOS 6 and iOS 7. This claim assumes that the iOS device is jailbroken, or that the examiner has access to the host computer that contains the pairing keys in escrow files. The tool is discussed in detail in *Chapter 6, iOS Forensic Tools*.

The following details explain the steps involved in obtaining a disk image from the iPhone 4S that has iOS 5 and is protected with a passcode in this example.

As a prerequisite, the iPhone 4S should already be jailbroken and OpenSSH is installed on it with the default root user password.

Set up the iPhone Data Protection Tools as explained in the previous sections. Edit `Makefile` in the `ramdisk_tools` folder, fix the iOS SDK version, and run the `make` command:

```
$cd iphone-dataprotection
$cd ramdisk_tools
$sudo make
```

Connect the iPhone to the computer via USB and establish the communication by running the `tcprelay.py` script as follows:

```
$cd iphone-dataprotection
$python usbmuxd-python-client/tcprelay.py -t 22:2222
```

Dump the iPhone user data partition using the following command:

```
$ssh root@127.0.0.1 "dd if=/dev/rdisk0s1s2  bs=8192" > data.dmg
```

Enter `alpine` as the password and hit *Enter*.

Download `kernel_patcher` from `https://code.google.com/p/iphone-dataprotection/issues/detail?id=49&q=a5` and move it to the `ramdisk_tools` folder with the following command:

```
$mv ~/Downloads/kernel_patcher ~/Documents/iphone-dataprotection/
```

Copy `kernel_patcher`, `bruteforce`, and `device_infos` scripts to the iPhone using the `scp` command:

```
$cd ramdisk_tools
$scp  -P 2222 kernel_patcher device_infos bruteforce
root@127.0.0.1:/var/root/
```

Enter `alpine` as the password and hit *Enter*.

Run the `ssh` command and grant execute permissions to the uploaded scripts with the following:

```
$ssh root@127.0.0.1 -p 2222
```

Enter `alpine` as the password and hit *Enter*:

```
iPhone# chmod +x kernel_patcher bruteforce device_infos
```

Run the `kernel_patcher` and `bruteforce` scripts. It patches the kernel, brute forces the passcode, decrypts the System keybag, and creates a plist file on the iPhone root directory, as shown in the following command lines:

```
iPhone#./ kernel_patcher
iPhone#./bruteforce
Writing results to f04d282cc7182d47.plist
[...]
```

Copy the plist file from the iPhone to the desktop using the `scp` command:

```
$scp -P 2222 root@127.0.0.1:/var/root/f04d282cc7182d47.plist .
```

To decrypt the disk image, run `emf_decrypter.py`, as follows:

```
$sudo python python_scripts/emf_decrypter.py data.dmg
f04d282cc7182d47.plist
```

Now, you should be able to fully examine the artifacts on the data partition.

Summary

The first step in the iPhone forensic examination is to acquire the data from the device. There are different ways to acquire data from an iPhone. This chapter covered physical acquisition techniques and techniques to bypass passcodes and data encryptions using open source methods. Physical acquisition is preferred as it recovers more data from the device; however, it is not possible to perform physical acquisition on all iOS devices. The following table summarizes the physical acquisition possibilities on iOS devices:

Model	Physical acquisition
iPhone 3G, 3GS, 4 iPad 1 iPod touch 2G, 3G, 4G	Yes (if no/easy passcode)
iPhone 4S, 5 iPad 2, 3, 4 and iPad mini iPod touch 5G	Only if jailbroken, and until iOS 6.1.2 (if no/easy passcode)
iPhone 5S and 5C	No

While physical acquisition is the best method for forensically obtaining the majority of the data from iOS devices, logical or backup files may exist or be the only method to extract data from the device. The next chapter discusses iOS device backup files in detail to include user, forensic, encrypted, and iCloud backup files and the methods to conduct your forensic examination.

4

Data Acquisition from iOS Backups

The physical acquisition of an iPhone provides the most data in an investigation, but you can also find a wealth of information on iPhone backups. iPhone users have several options to back up data present on their devices. iPhone users can choose to back up data to their computer using the **Apple iTunes** software or to the Apple cloud storage service known as **iCloud**. Every time an iPhone is synced with a computer or to iCloud, it creates a backup by copying the selected files from the device. The user can determine what is contained in the backup, so some may be more inclusive than others. Also, the user can back up to both a computer and iCloud, and the data derived from each location may differ. Sometimes, the best information available on an iOS device is recovered from a backup file.

In the previous chapter, we covered techniques to acquire data from an iPhone. This chapter covers **backup file acquisition** techniques using Apple's synchronization protocol from the device onto a computer or to iCloud. *Chapter 5, iOS Data Analysis and Recovery*, will then teach you how to analyze the data pulled from *Chapter 3, Data Acquisition from iOS Devices*, and *Chapter 4, Data Acquisition from iOS Backups*.

iTunes backup

A wealth of information is stored on any computer that has been previously synced with an iPhone. These computers, commonly referred to as host computers, can have historical data and passcode-bypass certificates. So, in a criminal investigation, a search warrant can be obtained to seize a computer that belongs to the suspect. iOS backup file forensics mainly involve analyzing an offline backup produced by an iPhone. However, the iTunes backup method is also useful in cases when physical acquisition of a device is not feasible. In this situation, examiners essentially create an iTunes backup of the device and analyze it using forensic software. Thus, it is important for an examiner to completely understand the backup process and the tools involved.

iPhone backup files can be created using the **iTunes** software, which is available for MAC OS X and Windows platforms. iTunes is a free utility provided by Apple for **data synchronization** and management between the iPhone and the computer. iTunes uses Apple's proprietary synchronization protocol to copy data from the iPhone to a computer. An iPhone can be synced with a computer using a USB or Wi-Fi. iTunes provides an option for encrypted backup, but by default it creates an unencrypted backup whenever an iPhone is synced. The backup copies of the iPhone can also be useful to recover data if the phone is lost or damaged.

iTunes is configured to automatically initiate the synchronization process once the iPhone is connected to the computer. To avoid unintended data exchange between the iPhone and the computer, disable the automatic synchronization process before connecting your iPhone to the forensic workstation. The following screenshot illustrates the option that disables automatic syncing in iTunes Version 11.1.3.

To disable auto-syncing in iTunes, perform the following steps:

1. Navigate to **iTunes | Preferences | Devices**.
2. Check **Prevent iPods, iPhones and iPads from syncing automatically** and click on the **OK** button.

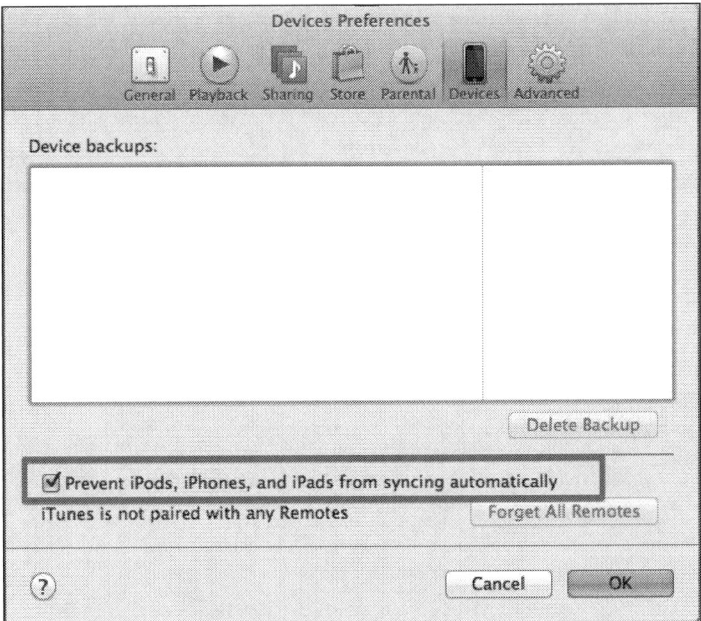

iTunes—disabling automatic sync

3. Once you verify the synchronization settings, connect the iPhone to the computer using a USB cable. If the connected iPhone is not protected with a passcode, iTunes immediately recognizes the device. This can be verified by the iPhone icon displayed on the upper-right corner of the iTunes interface as shown in the following screenshot:

4. If the connected iPhone is protected with a passcode, iTunes prompts the user to unlock the device before starting the sync process, as shown in the following screenshot. Once the iPhone is unlocked with a valid passcode, iTunes recognizes the device and allows the user to back up and sync with the computer. Once an iPhone is successfully synced with a computer, iTunes allows it to back up without unlocking the device when the same iPhone is connected to that computer again.

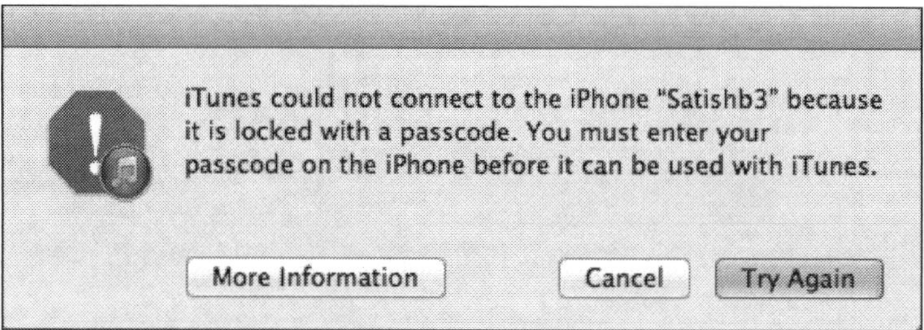

iTunes—iPhone locked message

5. Once iTunes recognizes the device, a single click on the iPhone icon displays the iPhone summary including the iPhone's name, capacity, firmware version, serial number, free space, and phone number, as shown in the following screenshot. The iPhone **Summary** page also displays the options to create backups.

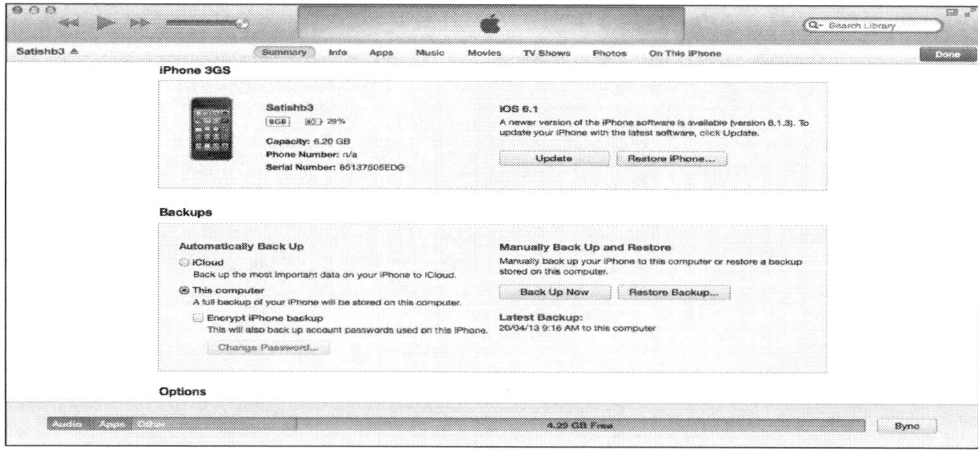

iTunes—iPhone summary

Pairing records

When iTunes detects the iPhone, sets of pairing records are exchanged between the iPhone and the computer. Pairing is the mechanism by which your computer establishes a trusted relationship with your device so that iTunes can communicate with it. Once a computer has been paired, it can access personal information on the device and can even initiate a backup of the device. Similar pairing occurs in iOS 7 with commercial forensic tools.

On the iPhone, pairing records are stored in the `/var/root/Library/Lockdown/pair_records/` directory. The directory will contain multiple pairing records if the device is paired with multiple computers. Pairing records are stored as a **property list** (`.plist`) file with a filename representing the unique identifier given to the computer. Property list files are binary formatted XML-like files, explained in detail in *Chapter 5, iOS Data Analysis and Recovery*. Pairing records on the device contain the HostID, root certificate, device certificate, and host certificate. For example, the content shown in the following screenshot was located in a pairing record on one particular iPhone with a file named `97D6299A-8EDA-454F-9C62-4BB031F45DD6.plist`. Pairing records stored on the iPhone are deleted only when the phone is restored to factory state.

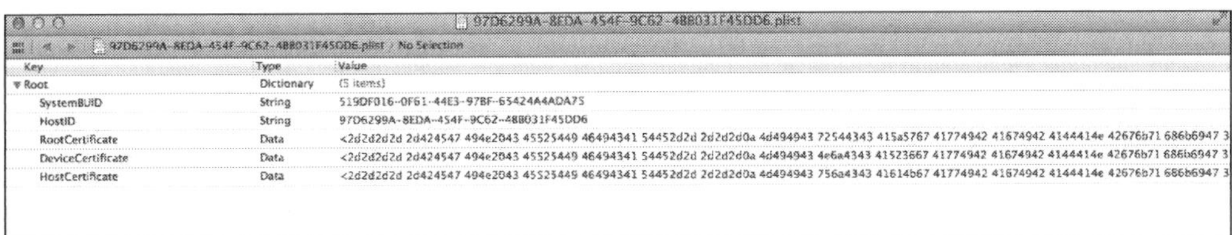

Pairing records on the iPhone

On the computer, pairing records are stored in a preconfigured location depending on the operating system as shown in the following table. Pairing records are stored as a property list file with a filename representing the iPhone's unique device identifier. Pairing records on the computer are known as **lockdown certificates**.

Operating system	Location
Windows	`%AllUserProfile%\Apple\Lockdown\`
Mac OS X	`/private/var/db/lockdown/`

Pairing records on the computer contain the device certificate, Escrow keybag, root certificate, host certificate, host private key, and root certificate and private key. For example, the content shown in the following screenshot was located in a pairing record on one particular computer with a file named `6c1b7aca59e2eba6f4635cfe7c4b2de1bd812898.plist`.

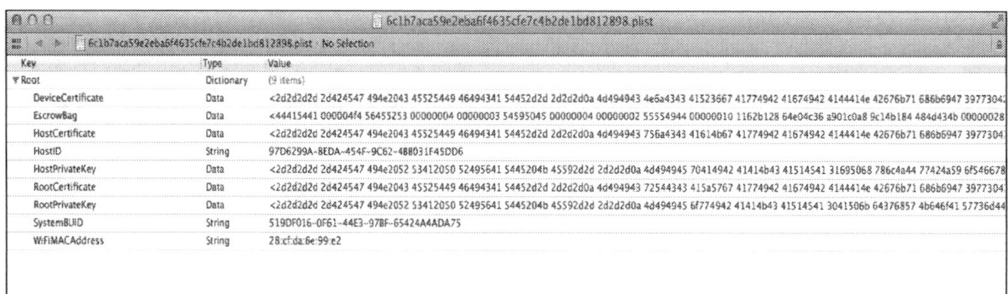

Pairing record on a computer

The Escrow keybag stored on the computer allows iTunes to back up and sync with the device even in a locked state. The Escrow keybag is a copy of the **System keybag** and contains a collection of data protection class keys that are used for encryption on the iPhone. Commercial tools that claim to be able to crack a locked iPhone without brute force require access to the host computer and thus, the Escrow keybag. The keybag improves the user experience during device synchronization and gives access to all classes of data on the device without entering the passcode.

The Escrow keybag is protected with a newly generated key computed from the key `0x835` and stored in an escrow record on the device. The escrow record is a property list file stored in the `/private/var/root/Library/Lockdown/escrow_records/` directory with a filename that represents the computer's unique identifier. Starting with iOS 5, escrow records are protected with the `UntilFirstUserAuthentication` data protection class, which ties the encryption to the user's passcode. So, the device passcode must be entered before backing up with iTunes for the first time.

Understanding the backup structure

When the iPhone is backed up to a computer, the backup files are stored in a backup directory, which exists as a 40-character hexadecimal string, and corresponds to the **Unique Device Identifier** (**UDID**) of the device. The backup process may take a considerable amount of time depending on the size of the data stored on the iPhone during the first backup. The location of the backup directory where your backup data is stored depends on the computer's operating system. The following table displays a list of the common operating systems and the default location of the iTunes backup directory:

Operating system	Backup directory location
Windows XP	`\Documents and Settings\[user name]\Application Data\Apple Computer\MobileSync\Backup\`
Windows Vista/7/8	`\Users\[user name]\AppData\Roaming\Apple Computer\MobileSync\Backup\`
Mac OS X	`~/Library/Application Support/MobileSync/Backup/` (~ represents your Home folder)

During the first sync, iTunes creates a backup directory and takes a complete backup of the device. On subsequent syncs, iTunes only backs up the files that are modified on the device and updates the existing backup directory. Also, when a device is updated or restored, iTunes automatically initiates a backup and takes a **differential backup**. A differential backup has the same name as the backup directory, but appended with a dash (-), the ISO date of the backup, a dash (-), and the time in a 24-hour format with seconds (`[UDID]+ '-' + [Date]+'-'+[Time stamp]`).

The iTunes backup makes a copy of everything on the device to include contacts, SMSes, photos, the calendar, music, call logs, configuration files, documents, the keychain, network settings, offline web application cache, bookmarks, cookies and application data, and so on. The backup also contains device details such as the serial number, UDID, SIM details, and phone number. This information can also be used to prove a relationship between the desktop and the mobile device.

The backup directory contains four standard files along with the individual data files, which may exist in various formats depending on the version of iTunes. Older versions will contain `*.mdbackup`, `*.mdata`, `*.mdinfo`, and some files with no file extensions. The standard files store details about the backup and the device from which it was derived. These file names are as follows:

- `info.plist`
- `manifest.plist`
- `status.plist`
- `manifest.mbdb`

The first three files are property list files that can be easily analyzed using the **Property List Editor** application on Mac OS X.

info.plist

The `info.plist` file stores details about the backed up device and typically contains the following information:

- **Device name and display name**: This is the name of the device, which typically includes the owner's name
- **ICCID**: This is the Integrated Circuit Card Identifier, which is the serial number of the SIM
- **Last backup date**: This is the timestamp of the last successful backup
- **IMEI**: This is the International Mobile Equipment Identity, which is used to uniquely identify the mobile phone
- **Phone Number**: This is the phone number of the device at the time of backup
- **Installed applications**: This is the list of application identifiers on the device
- **Product type and production version**: This is the device model and firmware version
- **Serial number**: This is the serial number of the device
- **iTunes version**: This is the version of iTunes that generated the backup
- **Target Identifier and Unique Identifier**: This is the UDID of the device

manifest.plist

The `manifest.plist` file describes the contents of the backup and typically contains the following information:

- **Applications**: This is a list of third-party applications installed on the backed up device, their version numbers, and bundle identifiers
- **Date**: This is the timestamp of a backup created or last updated
- **IsEncrypted**: This identifies whether the backup is encrypted or not. For encrypted backups the value is `True`, otherwise it is `False`
- **Lockdown**: This contains device details, last backup computer's name, and other remote syncing profiles
- **WasPasscodeSet**: This identifies whether a passcode was set on the device when it was last synced
- **Backup keybag**: Starting with iOS 4, a Backup keybag is created for each backup made by iTunes. The Backup keybag contains a new set of data protection class keys that are different from the keys in the System keybag, and backed up data is re-encrypted with the new class keys. Keys in the Backup keybag facilitate the storage of backups in a secure manner

status.plist

The `status.plist` file stores details about the backup status and typically contains the following information:

- **BackupState**: This identifies whether the backup is a new backup or one that has been updated
- **Date**: This is the timestamp of the last time the backup was modified
- **IsFullBackup**: This identifies whether or not the backup was a full backup of the device

manifest.mbdb

The `manifest.mbdb` file is a binary file and contains records about all other files in the backup directory along with the file sizes, file type, and file structure. The `manifest.mbdb` file header and record format are shown in the following tables.

Header

The file header is a fixed value of 6 bytes. This value acts as a magic string to identify the file format.

Type	Data	Description
uint8	`mbdb\5\0`	This files a magic string

The manifest.mbdb file header

Record

Each record in the `manifest.mbdb` file contains details about a file in the backup.

Type	Data	Description
String	Domain	This is the domain name.
String	Path	This is the file path.
String	Target	This is an absolute path for symbolic links.
String	Digest	This contains SHA1 hash `0xFF 0xFF` for directories and for AppDomain files, and `0x00 0x14` for SystemDomain files.
String	Encryption key	This indicates encrypted files and `0xFF 0xFF` for unencrypted files.
uint16	Mode	This identifies file type `0xA000` for symbolic link, `0x4000` for directory, and `0x8000` for regular files.

Type	Data	Description
uint64	inode number	This is a lookup entry in the inode table.
uint32	User ID	This is mostly `501`.
uint32	Group ID	This is mostly `501`.
uint32	Last modified time	This is the file's last modified time in the Unix time format.
uint32	Last accessed time	This is the file's last accessed time in the Unix time format.
uint32	Created time	This is the file created time in the Unix time format
uint64	Size	This is the length of a file. It is `0` for a symbolic link and a directory.
uint8	Protection class	This is the data protection class `0x1 To 0xB`.
uint8	Number of properties	This is the number of extended attributes.

The manifest.mbdb file record

Apart from the standard files, the backup directory also contains hundreds of backup files with varying file extensions depending on the version of iTunes used to create the backup, as described earlier. In the following screenshot, the backup was created with the latest version of iTunes in which the files do not contain a file extension. The backup files are uniquely named with a 40-character hexadecimal string. These filenames signify a unique identifier for each data set copied from the iPhone.

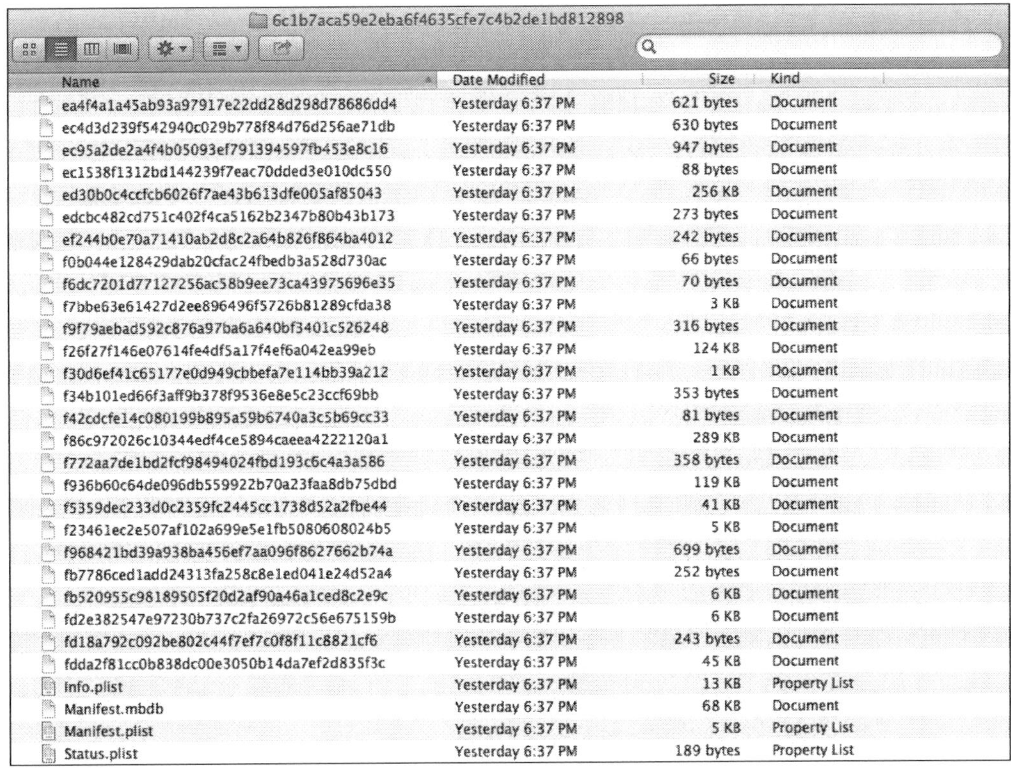

iPhone backup files

In iOS, files are categorized into 12 domains. All of the application files are classified into **AppDomain** and other files on the file system are classified into 11 system domains shown in the following screenshot. The list of system domains is stored in a property list file located under `/System/Library/Backup/Domains.plist` on the device.

The 40-character hexadecimal filename in the backup directory is the SHA1 hash value of the file path appended to the respective domain name with a dash (-) symbol.

For instance, the `AddressBook` database file is a member of **HomeDomain** and is located under `Library/AddressBook/AddressBook.sqlitedb`. The backup file name of `AddressBook` is `31bb7ba8914766d4ba40d6dfb6113c8b614be442`, which can be obtained by computing the SHA1 hash value of the following string: `HomeDomain-Library/AddressBook/AddressBook.sqlitedb`.

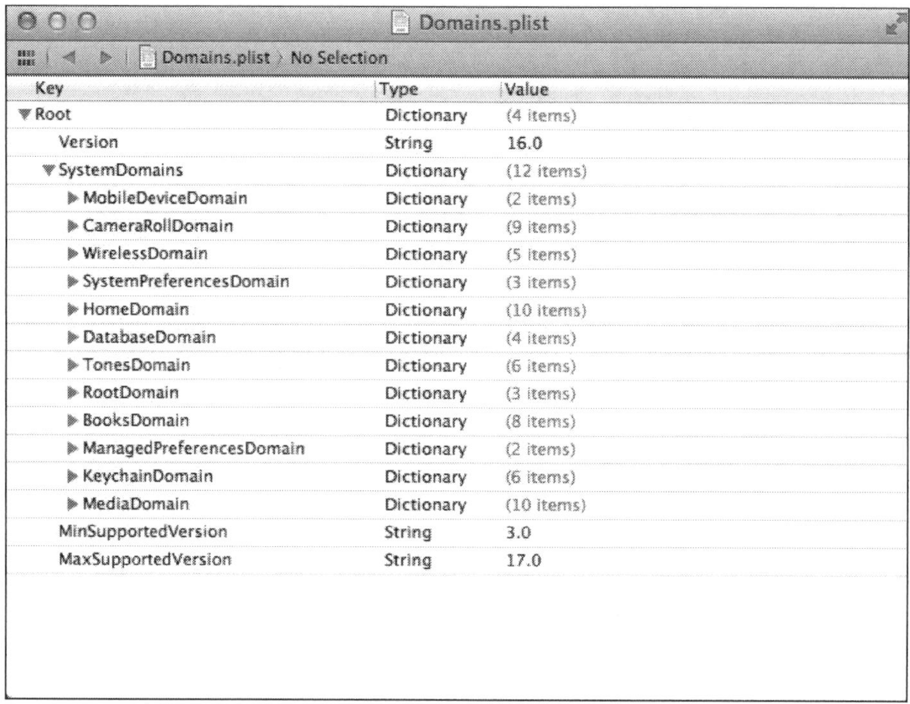

System domains on the iPhone

Unencrypted backup

To create an unencrypted backup, perform the following steps:

1. Connect the iPhone to the forensic workstation using a USB cable.
2. On the forensic workstation, launch iTunes.
3. Click on the iPhone icon displayed in the upper-right corner of the iTunes interface. It displays the iPhone **Summary** page.
4. In the iPhone summary page, select the **This computer** checkbox and click on the **Back Up Now** button.

Extracting unencrypted backups

There are many free tools available to analyze data from unencrypted backups. These tools parse the `manifest.mbdb` file, restore the filenames, and create the file structure that users see on the iPhone. Some of the popular tools include iPhone Backup Extractor, iPhone Backup Browser, and iPhone Data Protection Tools.

iPhone Backup Extractor

iPhone Backup Extractor is a free tool for Mac OS X, which can be downloaded from `http://supercrazyawesome.com/`. The backup extractor expects backup files to be located in the default location `~/Library/Application Support/MobileSync/Backup/`. So, you will need to copy any backups you wish to extract to the default location. iPhone Backup Extractor is a very easy tool to use.

To extract the backup, follow these steps:

1. Launch the app and click on the **Read Backups** button. It displays a list of backups available on the forensic workstation. Select the backup that you wish to extract and click on the **Choose** button, as shown in the following screenshot:

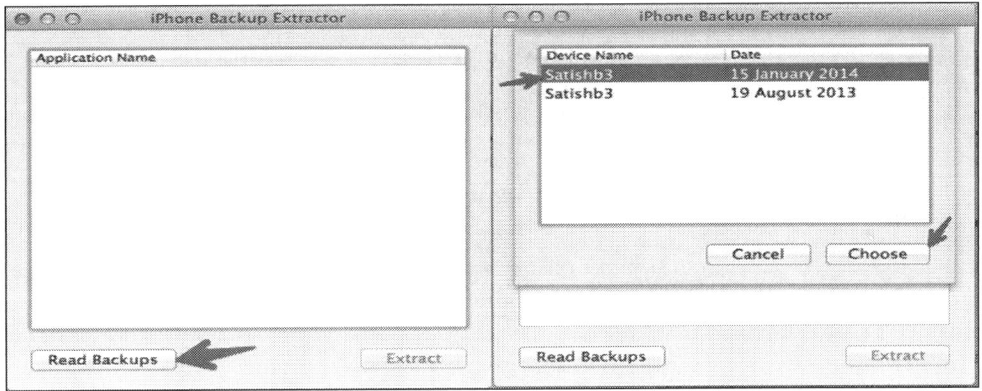

iPhone Backup Extractor—choosing backups

2. When you choose the backup, iPhone Backup Extractor allows you to extract the individual applications and the iOS file system backup, as shown in the following screenshot:

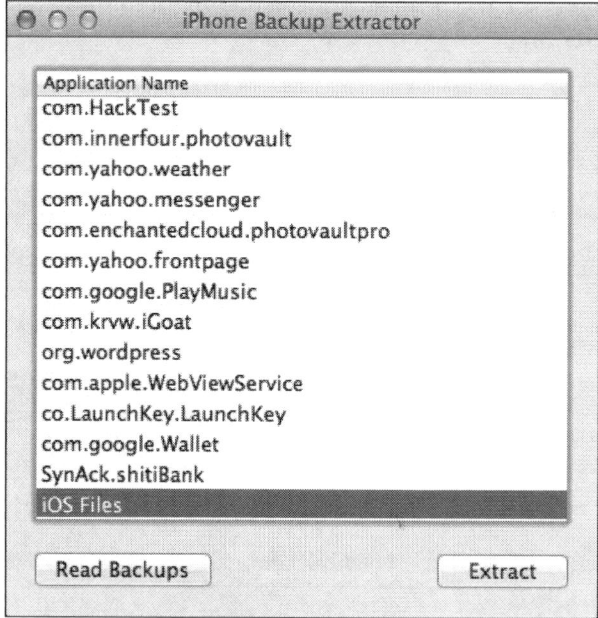

iPhone Backup Extractor

3. Choose the files you would like to extract and then click on **Extract**. It prompts for a destination directory to save the extracted files.

iPhone Backup Browser

iPhone Backup Browser is a free tool for Windows and can be downloaded from `http://code.google.com/p/iphonebackupbrowser/`. The tool requires **Microsoft .NET Framework 4** and Visual C++ 2010 runtime to be installed on the forensic workstation. The backup browser expects backup files to be located in the default location as mentioned in the preceding table. iPhone Backup Browser provides a GUI to view the backup data, as shown in the following screenshot:

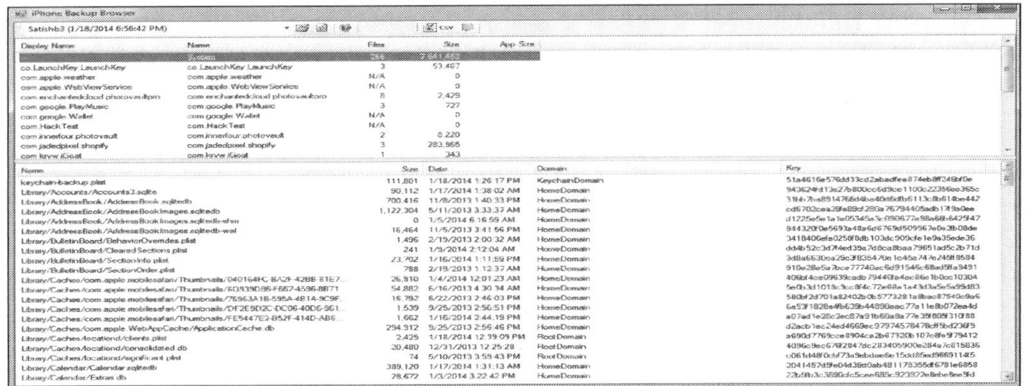

iPhone Backup Browser

iPhone Data Protection Tools

iPhone Data Protection Tools, an open source iOS forensic toolkit, can also be used to extract the backup files. To analyze data from the unencrypted backup file, set up iPhone Data Protection Tools as explained in *Chapter 3, Data Acquisition from iOS Devices*, and run the `backup_tool.py` script on your backup directory in a terminal window, as follows:

```
$cd iphone-dataprotection
$cd python_scripts
$sudo python backup_tool.py ~/Library/Application\
Support/MobileSync/Backup/6c1b7aca59e2eba6f4635cfe7c4b2de1bd812898/
Device Name : Satishb3
Display Name : Satishb3
Last Backup Date : 2014-01-07 12:58:13
IMEI : 012856001945212
Serial Number : 85137505EDG
Product Type : iPhone2,1
Product Version : 6.1
iTunes Version : 11.1.3
Extract backup to /Users/satishb3/Library/Application
Support/MobileSync/Backup/6c1b7aca59e2eba6f4635cfe7c4b2de1bd812898_ex
tract ? (y/n)
```

Type the letter `y` and hit *Enter*. The preceding script displays a number of messages indicating the current file being operated on, as shown in the following command lines:

```
Backup is not encrypted

Writing /Users/satishb3/Library/Application
Support/MobileSync/Backup_extract/HomeDomain/Library/Preferences/com.
apple.voiceservices.plist

Writing /Users/satishb3/Library/Application
Support/MobileSync/Backup_extract/CameraRollDomain/Media/DCIM/100APPL
E/IMG_0038.JPG

Writing /Users/satishb3/Library/Application
Support/MobileSync/Backup_extract/SystemPreferencesDomain/SystemConfi
guration/preferences.plist

Writing /Users/satishb3/Library/Application
Support/MobileSync/Backup_extract/HomeDomain/Library/Preferences/com.
apple.mobileipod.plist

[...]

Writing /Users/satishb3/Library/Application
Support/MobileSync/Backup/6c1b7aca59e2eba6f4635cfe7c4b2de1bd812898_ex
tract/HomeDomain/Library/Preferences/com.apple.springboard.plist
```

You can decrypt the keychain using the following command:

```
python keychain_tool.py -d "/Users/satishb3/Library/Application
Support/MobileSync/Backup/6c1b7aca59e2eba6f4635cfe7c4b2de1bd812898_ex
tract/KeychainDomain/keychain-backup.plist"
"/Users/satishb3/Library/Application
Support/MobileSync/Backup/6c1b7aca59e2eba6f4635cfe7c4b2de1bd812898_ex
tract/Manifest.plist"
```

The preceding script creates the `6c1b7aca59e2eba6f4635cfe7c4b2de1bd812898_extract` folder in the backup directory location and extracts the backup files into it by restoring the original filenames. The extracted backup files are stored in a number of domain directories as shown in the following screenshot. Now, you should be able to completely examine the artifacts on the backup files, which will be covered in detail in *Chapter 5, iOS Data Analysis and Recovery*. Pay attention to the directory names used in the command line as they vary for each device.

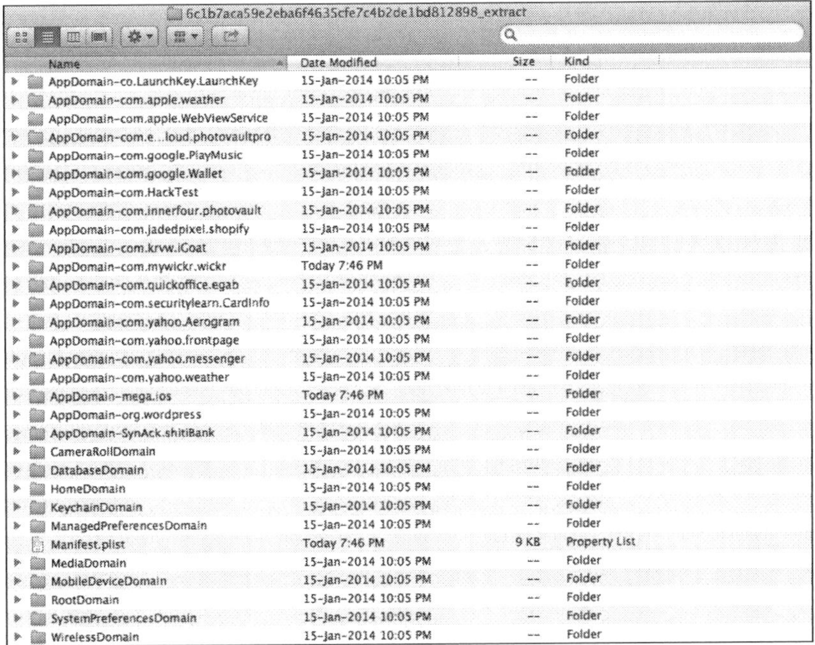

Extracted iPhone backup files

Decrypting the keychain

For unencrypted backups, all the backup files are stored unencrypted except the keychain. The keychain file contents are encrypted with a set of class keys in the Backup keybag. The Backup keybag itself is protected with a key (`0x835`) derived from the iPhone hardware key (UID key). So, in order to decrypt the keychain, you need to extract the key `0x835` from the device using the `demo_bruteforce.py` techniques explained in *Chapter 3, Data Acquisition from iOS Devices*.

The iPhone Data Protection tools also contain python scripts to decrypt the keychain file from the backup. To decrypt the keychain, run the following command in a terminal window and enter your device key `0x835` when prompted:

```
$sudo python keychain_tool.py -d "/Users/satishb3/Library/Application
Support/MobileSync/Backup/6c1b7aca59e2eba6f4635cfe7c4b2de1bd812898_ex
tract/KeychainDomain/keychain-backup.plist"
"/Users/satishb3/Library/Application
Support/MobileSync/Backup/6c1b7aca59e2eba6f4635cfe7c4b2de1bd812898_ex
tract/Manifest.plist"
```

```
This backup is not encrypted, without key 835 nothing in the keychain can
be decrypted
```

```
If you have key835 for device
6c1b7aca59e2eba6f4635cfe7c4b2de1bd812898 enter it (in hex)
33403aec43adea127459485bf5969502
```

The script extracts generic passwords, Internet passwords, certificates, and private keys from the keychain and displays them in a table as shown in the following screenshot:

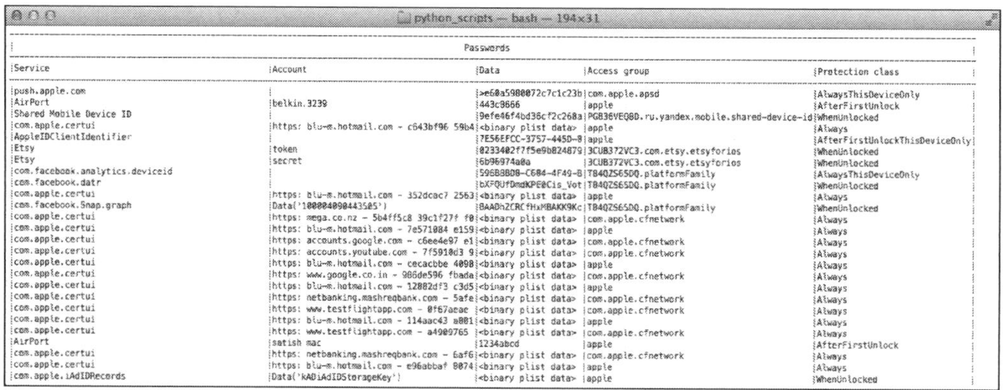

python_scripts — bash — 194×31

Passwords

Service	Account	Data	Access group	Protection class
push.apple.com		>e60a5980072c7c1c23b	com.apple.apsd	AlwaysThisDeviceOnly
AirPort	belkin.3239	443c9666	apple	AfterFirstUnlock
Shared Mobile Device ID		9efe46f4bd36cf2c268a	PGB36VEQ8D.ru.yandex.mobile.shared-device-id	WhenUnlocked
com.apple.certui	https: blu-m.hotmail.com - c643bf96 59b4	<binary plist data>	apple	Always
AppleIDClientIdentifier		7E56EFCC-3757-445D-8	apple	AfterFirstUnlockThisDeviceOnly
Etsy	token	8233402f7f5e9b824879	3CUB372VC3.com.etsy.etsyforios	WhenUnlocked
Etsy	secret	6b96974a0a	3CUB372VC3.com.etsy.etsyforios	WhenUnlocked
com.facebook.analytics.deviceid		596B8BD8-C684-4F49-8	T84QZS65DQ.platformFamily	AlwaysThisDeviceOnly
com.facebook.datr		bXFQUfDmdKPE0Cis_Vot	T84QZS65DQ.platformFamily	WhenUnlocked
com.apple.certui	https: blu-m.hotmail.com - 352dcac7 2563	<binary plist data>	apple	Always
com.facebook.Snap.graph	Data('100004090443505')	8AADhZCRCfHxMBAKK9Kc	T84QZS65DQ.platformFamily	WhenUnlocked
com.apple.certui	https: mega.co.nz - 5b4ff5c8 39c1f27f f0	<binary plist data>	com.apple.cfnetwork	Always
com.apple.certui	https: blu-m.hotmail.com - 7e571084 e159	<binary plist data>	apple	Always
com.apple.certui	https: accounts.google.com - c6ee4e97 e1	<binary plist data>	com.apple.cfnetwork	Always
com.apple.certui	https: accounts.youtube.com - 7f5910d3 9	<binary plist data>	com.apple.cfnetwork	Always
com.apple.certui	https: blu-m.hotmail.com - cecacbbe 4098	<binary plist data>	apple	Always
com.apple.certui	https: www.google.co.in - 986de596 fbada	<binary plist data>	com.apple.cfnetwork	Always
com.apple.certui	https: blu-m.hotmail.com - 12882df3 c3d5	<binary plist data>	apple	Always
com.apple.certui	https: netbanking.mashreqbank.com - 5afe	<binary plist data>	com.apple.cfnetwork	Always
com.apple.certui	https: www.testflightapp.com - 0f67aeae	<binary plist data>	com.apple.cfnetwork	Always
com.apple.certui	https: blu-m.hotmail.com - 114aac43 a001	<binary plist data>	apple	Always
com.apple.certui	https: www.testflightapp.com - a4909765	<binary plist data>	com.apple.cfnetwork	Always
AirPort	satish mac	1234abcd	apple	AfterFirstUnlock
com.apple.certui	https: netbanking.mashreqbank.com - 6af6	<binary plist data>	com.apple.cfnetwork	Always
com.apple.certui	https: blu-m.hotmail.com - e96abbaf 0074	<binary plist data>	apple	Always
com.apple.iAdIDRecords	Data('kADiAdIDStorageKey')	<binary plist data>	apple	WhenUnlocked

A decrypted keychain

Encrypted backup

iTunes provides an option for the users to encrypt their backups using a password. Forensic examiners may elect to create an encrypted backup to protect the evidence. It is pertinent that the examiner documents the password should this method be used.

To create an encrypted backup, perform the following steps:

1. Connect the iPhone to the forensic workstation using a USB cable.
2. On the forensic workstation, launch iTunes.
3. Click on the iPhone icon displayed in the upper-right corner of the iTunes interface. It displays the iPhone summary page.
4. In the iPhone summary page, select the **This computer** checkbox and select the **Encrypt iPhone backup** option. Selecting the option prompts you to enter a password, as shown in the following screenshot.
5. Set a password and click on the **Back Up Now** button. It creates an encrypted backup.

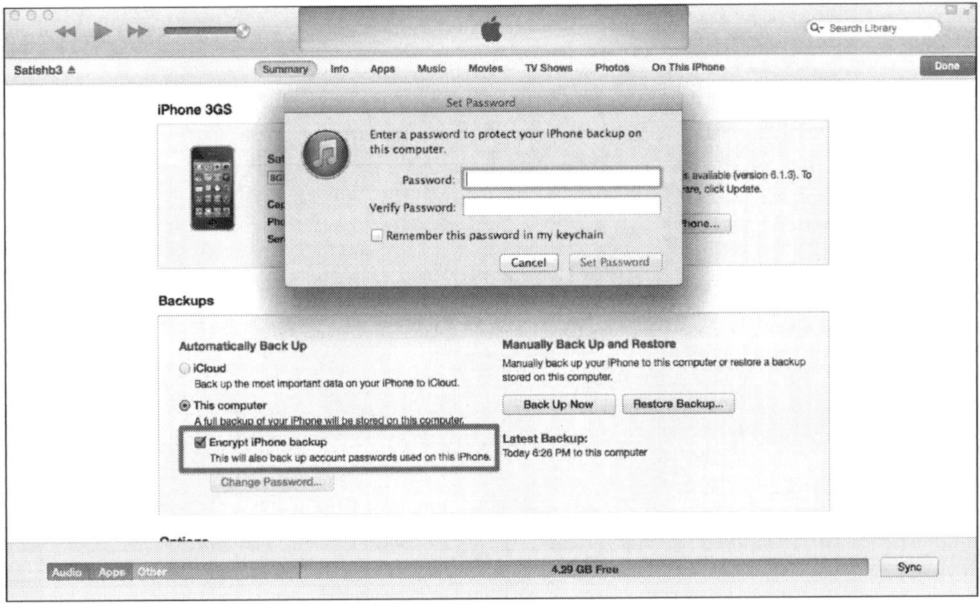

iTunes—encrypted backup

If a backup is password protected, the password is set on the device itself and stored in the keychain file. Also, whenever the device is connected to iTunes, it automatically chooses the **Encrypt iPhone backup** option regardless whether the users own a copy of iTunes being used on their computer or someone else's. So, even if you have access to the suspect's iPhone, you cannot produce an unencrypted backup unless you know the backup password.

Extracting encrypted backups

For encrypted backups, the backup files are encrypted using the `AES256` algorithm in the CBC mode, with a unique key and a null **IV (initialization vector)**. The unique file keys are protected with a set of class keys from the Backup keybag. The class keys in the Backup keybag are protected with a key derived from the password set in iTunes through 10,000 iterations of **PBKDF2 (Password-Based Key Derivation Function 2)**. Both open source and commercial tools provide support for an encrypted backup file parsing if the password is known. Some tools won't even prompt for a password, which make them useless in a forensic investigation. iPhone Data Protection Tools is capable of extracting data from encrypted backup files if the password is known.

iPhone Data Protection Tools

iPhone Data Protection Tools contains Python scripts to decrypt the backup when the backup password is available. To decrypt and acquire data from the encrypted backup, in a terminal window, run the `backup_tool.py` script on your backup directory and enter the backup password when prompted, as shown in the following commands:

```
$cd iphone-dataprotection

$cd python_scripts

$sudo python backup_tool.py ~/Library/Application\
Support/MobileSync/Backup/6c1b7aca59e2eba6f4635cfe7c4b2de1bd812898/

Device Name : Satishb3

Display Name : Satishb3

Last Backup Date : 2014-01-15 16:34:13

IMEI : 012856001945212

Serial Number : 85137505EDG

Product Type : iPhone2,1

Product Version : 6.1

iTunes Version : 11.1.3

Extract backup to /Users/satishb3/Library/Application
Support/MobileSync/Backup/6c1b7aca59e2eba6f4635cfe7c4b2de1bd812898_ex
tract ? (y/n)
```

Type the letter `y` and hit *Enter*. The script displays a number of messages indicating the current file being operated upon, as follows:

```
Backup is encrypted

Enter backup password:

12345

Writing /Users/satishb3/Library/Application
Support/MobileSync/Backup/6c1b7aca59e2eba6f4635cfe7c4b2de1bd812898_ex
tract/HomeDomain/Library/Preferences/com.apple.voiceservices.plist

Writing /Users/satishb3/Library/Application
Support/MobileSync/Backup/6c1b7aca59e2eba6f4635cfe7c4b2de1bd812898_ex
tract/CameraRollDomain/Media/DCIM/100APPLE/IMG_0038.JPG
```

```
Writing /Users/satishb3/Library/Application
Support/MobileSync/Backup/6c1b7aca59e2eba6f4635cfe7c4b2de1bd812898_ex
tract/SystemPreferencesDomain/SystemConfiguration/preferences.plist
[...]
Writing /Users/satishb3/Library/Application
Support/MobileSync/Backup/6c1b7aca59e2eba6f4635cfe7c4b2de1bd812898_ex
tract/HomeDomain/Library/Preferences/com.apple.springboard.plist
You can decrypt the keychain using the following command:
python keychain_tool.py -d "/Users/satishb3/Library/Application
Support/MobileSync/Backup/6c1b7aca59e2eba6f4635cfe7c4b2de1bd812898_ex
tract/KeychainDomain/keychain-backup.plist"
"/Users/satishb3/Library/Application
Support/MobileSync/Backup/6c1b7aca59e2eba6f4635cfe7c4b2de1bd812898_ex
tract/Manifest.plist"
```

The script creates the `6c1b7aca59e2eba6f4635cfe7c4b2de1bd812898_extract` folder in the backup directory location, then decrypts and extracts the backup files into a number of domain directories by restoring the original filenames.

Decrypting the keychain

Encrypted backup files can be cracked using brute force attacks in both the command line and GUI tools. For encrypted backups, the keychain items protected with the `ThisDeviceOnly` data protection class are encrypted using a set of class keys that are protected with the key `0x835`. All other keychain items are encrypted using a set of class keys that are protected with a password set in iTunes. If you want to extract the `ThisDeviceOnly` protected items, you need to extract a key `0x835` from the device using the `demo_bruteforce.py` techniques explained in *Chapter 3, Data Acquisition from iOS Devices.*

iPhone Data Protection Tools contain Python scripts to decrypt the keychain file from the encrypted backup. To decrypt the keychain, run the following command in a terminal window and enter the backup password when prompted. The script also prompts to enter the key `0x835`; press *Enter* if you don't have the key `0x835`.

```
$sudo python keychain_tool.py -d "/Users/satishb3/Library/Application
Support/MobileSync/Backup/6c1b7aca59e2eba6f4635cfe7c4b2de1bd812898_ex
tract/KeychainDomain/keychain-backup.plist"
"/Users/satishb3/Library/Application
Support/MobileSync/Backup/6c1b7aca59e2eba6f4635cfe7c4b2de1bd812898_ex
tract/Manifest.plist"
```

Once completed, the script extracts generic passwords, Internet passwords, and certificates and private keys from the keychain, and displays them in a table.

iPhone Password Breaker

iPhone Password Breaker is a GPU-accelerated commercial tool from Elcomsoft developed for the Windows platform. The tool can decrypt the encrypted backup file when the backup password is not available. The tool provides an option to launch a password brute-force attack on the encrypted backup if the backup password is not available. iPhone Password Breaker tries to recover the plain-text password that protects the encrypted backup using dictionary and brute force attacks. Passwords, which are relatively short and simple, can be recovered in a reasonable time. But if the backup is protected with a strong and complex password, breaking it can take forever.

To brute force the backup password, perform the following steps:

1. Launch the iPhone Password Breaker tool and the tool's main screen will appear, as shown in the following screenshot.
2. Navigate to **File | Open | Backup**. A list of available device backups is displayed and a lock symbol is shown next to the encrypted device backups, as shown in the following screenshot:

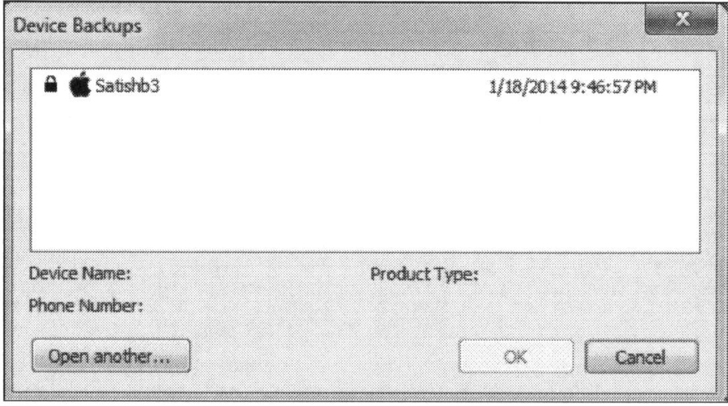

iPhone Password Breaker - Choose backup

3. Configure the brute-force pattern in the **Attacks** section and click on the **Start** button to start the brute force attack. If the brute force attack is successful, the tool displays the password on the main screen, as shown in the following screenshot:

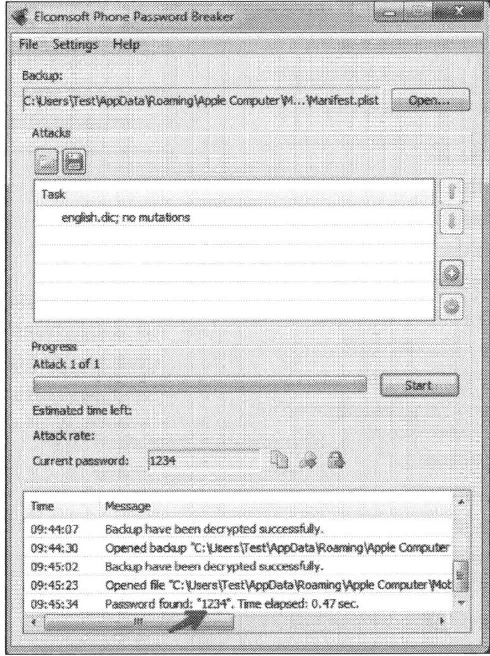

iPhone Password Breaker – password brute force

iCloud backup

iCloud is a cloud storage and cloud computing service by Apple launched in October 2011. The service allows users to keep data such as calendars, contacts, reminders, photos, documents, bookmarks, applications, notes, and more in sync across multiple compatible devices (iOS devices running with iOS 5 or later, computers with Mac OS X 10.7.2 or later, and Microsoft Windows) using a centralized iCloud account. The service also allows users to wirelessly and automatically back up their iOS devices to iCloud. iCloud also provides other services such as **Find My iPhone**—to track a lost phone and wipe it remotely, **Find My Friends**—to share location with friends and notify the user when a device arrives at a certain location, and so on.

Signing up with iCloud is free and simple to do with an Apple ID. When you sign up for iCloud, Apple grants you access to 5 GB of free remote storage. If you need more storage, you can purchase the upgrade plan. To keep your data secure, Apple enforces users to choose a strong password when creating an Apple ID to use with iCloud. The password must have a minimum of eight characters, a number, an uppercase letter, and a lowercase letter.

iOS devices running on iOS 5 and later allow users to back up the device settings and data to iCloud. Data backed up includes photos, videos, documents, application data, device settings, messages, contacts, calendar, e-mail, keychain, and so on. You can turn on iCloud backup on your device by navigating to **Settings** | **iCloud** | **Storage & Backup**, as shown in the following screenshot. iCloud can automatically back up your data when your phone is plugged in, locked, and connected to Wi-Fi. This is to say, iCloud backups represent a fresh and near real-time copy of information stored on the device.

iCloud backup toggle on the iPhone

You can also initiate an iCloud backup from a computer by connecting the device to iTunes and choosing the iCloud option. iCloud backups are incremental, that is, once the initial iCloud backup is completed, all the subsequent backups only copy the files that are changed on the device. iCloud secures your data by encrypting it when it is transmitted over the Internet, storing it in an encrypted format on the server, and using secure tokens for authentication.

[iCloud does not encrypt **Email** and **Notes** stored on the server to be consistent with standard industry practices.]

Apple's built-in apps (for example, **Email** and **Contacts**) use a secure token to access iCloud services. Use of secure tokens for authentication eliminates the need to store the iCloud password on devices and computers.

Extracting iCloud backups

Online backups stored on the iCloud are commonly retrieved when the original iPhone is damaged or lost. To extract a backup from iCloud, you must know the user's Apple ID and password. With the known Apple ID and password, you can log on to `www.icloud.com` and get access to contacts, notes, e-mail, calendar, photos, reminders, and more. To extract the complete backup from iCloud, you can use Elcomsoft iPhone Password Breaker. As iCloud is not the fastest cloud storage, downloading a large backup with iPhone Password Breaker can take hours. To speed up the investigation, the tool provides an option to download the selected files.

To extract the iCloud backup, perform the following steps:

1. Launch the iPhone Password Breaker.
2. Navigate to **File | Apple | Get Backup from iCloud**. It displays a prompt to sign in with your Apple ID, as shown in the following screenshot:

3. Successfully signing in with your Apple ID lists the available device backups, as shown in the following screenshot:

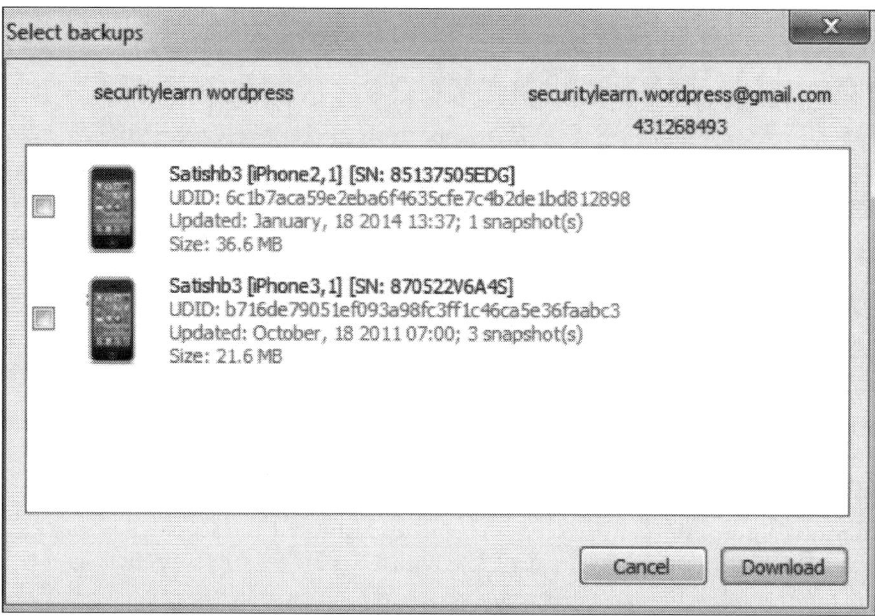

4. Select the backup you need and click on **Download**. It prompts you for a destination directory to save the extracted files into a number of domain directories by restoring the original filenames. The tool also provides an option to download the backup without restoring the original filenames so that you can use third-party software for analysis.

For iCloud backups, the keychain file contents are encrypted with a set of class keys in the Backup keybag. The Backup keybag itself is protected with a key (`0x835`) derived from the iPhone hardware key (UID key). You can follow the techniques explained in the preceding sections to decrypt the keychain from the extracted iCloud backup.

Summary

iPhone backups contain essential information that may be your only source of evidence for the iPhone. Information stored in iPhone backups includes photos, videos, contacts, e-mail, call logs, user accounts and passwords, applications, device settings, and so on. This chapter covered techniques to create backup files and retrieve data from iTunes and iCloud backups including encrypted backup files, wherever possible. *Chapter 5, iOS Data Analysis and Recovery*, goes further into the forensic investigation by showing the examiner how to analyze the data recovered from the backup files. Areas containing data of potential evidentiary value will be explained in detail.

5
iOS Data Analysis and Recovery

A key aspect in iPhone forensics is to examine and analyze the data acquired from an iPhone to interpret the evidence. Data on most iOS devices is encrypted and requires that the data partition be decrypted prior to an examination. In the previous chapters, you learned various techniques to acquire data from an iPhone. The raw disk image obtained during physical acquisition, the file system dump or the logical/ backup file contains hundreds of data files. This chapter will help you to understand how data is stored on the iPhone and will walk you through the important files in order to recover the most data possible.

Timestamps

Before examining the data, it is important to understand the different timestamps used on the iPhone. Timestamps found on the iPhone are presented either in the Unix timestamp or Mac absolute time format. The examiner must ensure that the tools properly convert the timestamps for the files. Access to the raw SQLite files will allow the examiner to verify the timestamps manually.

Unix timestamps

A Unix timestamp is the number of seconds that offsets the **Unix epoch** time, which starts on January 1, 1970. A Unix timestamp can be converted easily using the `date` command on a Mac workstation or using an online Unix epoch convertor on a Windows workstation. The `date` command is shown as follows:

```
$date -r 1388538061
Wed Jan 1 06:31:01 IST 2014
```

Mac absolute time

iOS devices adopted the use of **Mac absolute time** with iOS 5 for most of the data. Mac absolute time is the number of seconds that offsets the Mac epoch time, which starts on January 1, 2001. The difference between the Unix epoch time and the Mac epoch time is exactly 978,307,200 seconds. To convert the Unix epoch time to Mac absolute time, add 978,307,200 to it and calculate it as a Unix timestamp. For example, the `date` command can be used to covert Mac absolute time is shown as follows:

```
$date -r `echo '389894124 + 978307200'| bc`
Fri May 10 21:25:24 IST 2013
```

Online converters prove to be useful to convert both Mac epoch and Unix timestamps for iOS devices.

SQLite databases

SQLite is an open source, in-process library that implements a self-contained, zero configuration, and transactional SQL database engine. It's a complete database with multiple tables, triggers, and views that are contained in a single cross-platform file. As SQLite is portable, reliable, and small, it is a popular database format that appears in many mobile platforms.

Apple iOS devices, like other smartphones, make heavy use of SQLite databases for data storage. Many of the built-in applications such as Phone, Messages, Mail, Calendar, and Notes store data in SQLite databases. Apart from that, third-party applications installed on the device also leverage SQLite databases for data storage.

SQLite databases are created with or without a file extension. They typically have `.sqlitedb` or `.db` file extensions, but some databases are given other extensions as well. Data in SQLite files is broken up into tables that contain the actual data. To access the data stored in these files, you need a tool that can read them. Some good free tools are:

- SQLite Browser, which can be downloaded from `https://github.com/rp-/sqlitebrowser`.
- SQLite command-line client, which you can download from `http://www.sqlite.org/`.
- SQLite Professional (`https://www.sqlitepro.com/`), a free **graphical user interface** (**GUI**) from Hankinsoft Development for Mac OS X users. You can download it from Mac's App Store.
- SQLite Spy, a free GUI tool for Windows. You can download it from `http://www.yunqa.de/delphi/doku.php/products/sqlitespy/index`.

Mac OS X includes the SQLite command-line utility (**sqlite3**) by default. This command-line utility can easily access individual files and issue SQL queries against a database. So, in the following sections we will use the `sqlite3` command-line utility to retrieve data from various SQLite databases. Before retrieving the data, the basic commands you will need to learn are explained in the following sections:

Connecting to a database

Manual examination of iOS SQLite database files is possible with the use of free tools. The following is an example of how to examine a database using native Mac commands in the terminal. Make sure your device image is mounted as read-only to prevent changes being made to the original evidence. To connect to a SQLite database from the command line, run the `sqlite3` command in the terminal by entering your database file. This will give you a SQL prompt where you can issue SQL queries:

```
$sqlite3 filename.sqlitedb
SQLite version 3.7.12 2012-04-03 19:43:07
Enter ".help" for instructions
Enter SQL statements terminated with a ";"
sqlite>
```

To disconnect, use the `.exit` command. It exits the SQLite client and returns to the terminal prompt.

SQLite special commands

Once you connect to a database, there are a number of built-in SQLite commands known as **dot commands** that can be used to obtain information from the database files. You can obtain the list of special commands by issuing the `.help` command in the SQLite prompt. These are SQLite-specific commands and do not require a semicolon at the end of the command. Most commonly used dot commands include the following:

- `.tables`: This lists all of the tables within a database. The following example displays the list of tables found inside the `sms.db` database:

  ```
  sqlite> .tables
  _SqliteDatabaseProperties  chat_message_join
  attachment                 handle
  chat                       message
  chat_handle_join           message_attachment_join
  ```

- `.schema table-name`: This displays the SQL `CREATE` statement used to construct the table. The following example displays the schema for the handle table, which is found inside the `sms.db` database:

  ```
  sqlite> .schema handle

  CREATE TABLE handle ( ROWID INTEGER PRIMARY KEY
  AUTOINCREMENT UNIQUE, id TEXT NOT NULL, country TEXT,
  service TEXT NOT NULL, uncanonicalized_id TEXT, UNIQUE
  (id,service) );
  ```

- `.dump table-name`: This dumps the entire content of a table into SQL statements. The following example displays the dump of the `handle` table, which is found inside the `sms.db` database:

  ```
  sqlite> .dump handle

  PRAGMA foreign_keys=OFF;

  BEGIN TRANSACTION;

  CREATE TABLE handle ( ROWID INTEGER PRIMARY KEY
  AUTOINCREMENT UNIQUE, id TEXT NOT NULL, country TEXT,
  service TEXT NOT NULL, uncanonicalized_id TEXT, UNIQUE
  (id,service) );

  INSERT INTO "handle"
  VALUES(7,'9951512182','in','SMS','9908923323');

  COMMIT;
  ```

- `.output file-name`: This redirects the output to a file on the disk instead of showing it on the screen.
- `.headers on`: This displays the column title whenever you issue a `SELECT` statement.
- `.help`: This displays the list of available SQLite dot commands.
- `.exit`: This disconnects from the database and exits the SQLite command shell.
- `.mode MODE`: This sets the output mode where `MODE` can be csv, HTML, tabs, and so on.

Make sure there is no space in between the SQLite prompt and the dot command, otherwise the entire command will be ignored.

Standard SQL queries

In addition to the SQLite dot commands, standard SQL queries such as `SELECT`, `INSERT`, `ALTER`, `DELETE`, and more can be issued to SQLite databases on the command line. Unlike the SQLite dot commands, the standard SQL queries expect a semicolon at the end of the command.

Most of the databases you will examine will contain only a reasonable number of records, so you can issue a `SELECT` statement, which outputs all of the data contained in the table. The following example displays the values in the `handle` table, which is found inside the `sms.db` database:

```
sqlite> select * from handle limit 1;
7|9951512182|in|SMS|9908923323
```

Important database files

Raw disk images, file system dumps the backup that you extracted as per the instructions in *Chapter 3, Data Acquisition from iOS Devices*, and *Chapter 4, Data Acquisition from iOS Backups*, will contain the following SQLite databases that may be important to your investigation. The files shown in the following sections are extracted from an iOS 6 device. As Apple adds new features to the built-in applications with every iOS release, the format of the files may vary for different iOS versions. So, you may need to modify the queries listed slightly to work on your iOS version. More information regarding important database files can be found at `http://www.zdziarski.com/blog/wp-content/uploads/2013/05/iOS-Forensic-Investigative-Methods.pdf`.

Address book contacts

The address book contains a wealth of information about the owner's personal contacts. With the exception of third-party applications, the address book contains contact entries for all of the contacts stored on the device. The address book database is a `HomeDomain` file and can be found at `private/var/mobile/Library/AddressBook/AddressBook.sqlitedb`.

`AddressBook.sqlitedb` contains several tables, of which three are of particular interest:

- `ABPerson`: This contains the name, organization, notes, and more for each contact.

- `ABMultiValue`: This contains phone numbers, e-mail addresses, website URLs, and more for the entries in the `ABPerson` table. The `ABMultiValue` table uses a `record_id` file to associate the contact information with a `rowid` from the `ABPerson` table.
- `ABMultiValueLabel`: This table contains labels to identify the kind of information stored in the `ABMultiValue` table.

Some of the data stored within the `AddressBook.sqlitedb` file could be from third-party applications. The examiner should manually examine the application file folders to ensure that all the contacts are accounted for and examined.

You can run the following commands to dump the address book into a CSV file named `AddressBook.csv`:

```
$sqlite3 AddressBook.sqlitedb
SQLite version 3.7.12 2012-04-03 19:43:07
Enter ".help" for instructions
Enter SQL statements terminated with a ";"
sqlite>.mode csv
sqlite>.output AddressBook.csv
sqlite>.headers on
sqlite> SELECT
        p.rowid, p.first, p.middle, p.last,
        datetime(p.creationDate+978307200,'unixepoch') as
        creationdate,
        case when m.label in
           (SELECT rowid from ABMultiValueLabel)
        then
           (SELECT value from ABMultiValueLabel where
m.label=rowid)
        else
           m.label end as Type,
        m.value, p.organization, p.department,
        p.note, p.birthday, p.nickname, p.jobtitle,
        datetime(p.modificationDate + 978307200, 'unixepoch') as
        modificationdate
      FROM ABPerson p,ABMultiValue m
      WHERE p.rowid=m.record_id and m.value not null
      ORDER by p.rowid ASC;
sqlite>.exit
```

The preceding query cross-references the data across the three tables and retrieves the contact information stored in the database. The query also converts the Mac absolute time into a readable form using the SQLite `datetime` function.

Address book images

In addition to the address book's data, each contact may contain an image associated with it. This image is displayed on the screen whenever the user receives an incoming call from a particular contact. The address book images database is a `HomeDomain` file and can be found at `/private/var/mobile/Library/AddressBook/AddressBookImages.sqlitedb`.

The `ABFullSizeImage` table in the `AddressBookImages.sqlitedb` file contains images in binary data. To extract the images, use SQLite's `.output` and `.dump` commands to create a text file and dump the database into this file in a SQL text format, as shown in the following command lines:

```
$sqlite3 AddressBookImages.sqlitedb
SQLite version 3.7.12 2012-04-03 19:43:07
Enter ".help" for instructions
Enter SQL statements terminated with a ";"
sqlite> .output AddressBookImages.txt
sqlite> .dump ABFullSizeImage
sqlite> .exit
```

The text file contains the image data in a hexadecimal encoding format. To convert this output back to binary data and grab the images, run the `AddressBookImageGrabber.py` Python script on the dump file, as shown in the following command. The Python script source code is available in the code bundle of the book.

```
$Python AddressBookImageGrabber.py AddressBookImages.txt
Writing ./AddressBookImages-Output/397.jpeg
Writing ./AddressBookImages-Output/129.jpeg
Writing ./AddressBookImages-Output/73.jpeg
Writing ./AddressBookImages-Output/508.jpeg
[...]
Writing ./AddressBookImages-Output/456.jpeg
Writing ./AddressBookImages-Output/141.jpeg
Total 93 images are extracted
```

Downloading the example code

You can download the example code files for all Packt books you have purchased from your account at http://www.packtpub.com. If you purchased this book elsewhere, you can visit http://www.packtpub.com/support and register to have the files e-mailed directly to you.

The script will create a directory named AddressBookImages-Output and place the extracted JPEG images onto it. The images can be viewed using a standard image viewer.

The filename of each image will be the record identifier, which is associated with the AddressBook.sqlite database so that you can associate each image with a contact.

Make sure you are using Python 2.7 to run the Python scripts.

Call history

Phone or FaceTime calls placed, missed, and received by the user are logged in the call history, along with other metadata such as call duration, date/time, and more. This could be of interest to an examiner. The call history database is a WirelessDomain file and can be found at /private/var/wireless/Library/CallHistory/call_history.db. The database contains a maximum of 100 calls listed as active messages. Any calls placed, missed, or received above 100 will be stored in the database and the oldest record will be removed. However, this data will remain in the SQLite free pages and can be recovered through manual hex examination.

The Call table in the call_history.db file contains the call history. Each record in the call table indicates the phone number of a remote party, a UNIX timestamp of when the call was initiated, the duration of the call in seconds, a status flag to identify whether the call was an outgoing call (flag 5), incoming call (flag 4), blocked call (flag 8), or FaceTime call (flag 16), an identifier that is associated with the address book contacts (-1 for unknown contact), the mobile county code (MCC), and the mobile network code (MNC). You can find a list of MCC/MNC codes at http://en.wikipedia.org/wiki/Mobile_country_code.

FaceTime status flags may vary depending on the method used to initiate the call. For example, data plans utilize different flags than Wi-Fi calls. If the status flag starts with a 2, it is likely to be a Wi-Fi initiated call. If it starts with a 1, as defined earlier, it represents a FaceTime call initiated with a data plan on the device. There are several status flags available for FaceTime calls and these vary between iOS devices.

You can run the following commands to dump the call history into a CSV file named `callhistory.csv`:

```
$sqlite3 call_history.db
SQLite version 3.7.12 2012-04-03 19:43:07
Enter ".help" for instructions
Enter SQL statements terminated with a ";"
sqlite>.mode csv
sqlite>.output callhistory.csv
sqlite>.headers on
sqlite> SELECT
          rowid, address,
          datetime(date,'unixepoch','localtime') as date,
          duration || " sec" as duration,
          case flags
             when 4 then "Incoming"
             when 5 then "Outgoing"
             when 8 then "Blocked"
             when 16 then "Facetime"
             else "Dropped"
          end as flags,
          id, country_code, network_code
       FROM call
       ORDER BY rowid ASC;
sqlite>.exit
```

SMS messages

The **Short Message Service** (**SMS**) database contains text and multimedia messages that were sent from and received by the device, along with the phone number of the remote party, date and time, and other carrier information. Starting with iOS 5, iMessages data is also stored in the SMS database. iMessage allows users to send SMS and MMS messages over a cellular or Wi-Fi network to other iOS or OS X users, thus providing an alternative to SMS. The SMS database is a `HomeDomain` file and can be found at `/private/var/mobile/Library/SMS/sms.db`.

You can run the following commands to dump the SMS database into a CSV file named `sms.csv`:

```
$sqlite3 sms.db
SQLite version 3.7.12 2012-04-03 19:43:07
```

```
Enter ".help" for instructions
Enter SQL statements terminated with a ";"
sqlite>.mode csv
sqlite>.output sms.csv
sqlite>.headers on
sqlite> SELECT
         m.rowid as rowid,
         datetime(date + 978307200, 'unixepoch') as date,
         h.id as "phone number", m.service as service,
         case is_from_me
             when 0 then "Received"
             when 1 then "Sent"
             else "Unknown"
         end as type,
         case
             when date_read > 0
                 then datetime(date_read+978307200,'unixepoch')
             when date_delivered > 0
                 then datetime(date_delivered+978307200,'unixepoch')
            else NULL
          end as "Date Read/Sent", text
       FROM message m, handle h
       WHERE h.rowid = m.handle_id
       ORDER BY m.rowid ASC;
sqlite>.exit
```

SMS Spotlight cache

Spotlight is a device-wide search feature, which allows the user to search across all the applications on the device. The SMS data is indexed and stored in a database for a quick search. The SMS Spotlight cache database is a `HomeDomain` file and can be found at `/private/var/mobile/Library/Spotlight/com.apple.MobileSMS/SMSSeaerchIndex.sqlite`. The file contains both active and deleted SMS messages. The following screenshot is an example of the output as viewed in SQLite Browser. This is a great place to recover SMS messages that are no longer present in the SMS database file. Note that the SMS Spotlight cache filename may vary depending on the version of the iOS device.

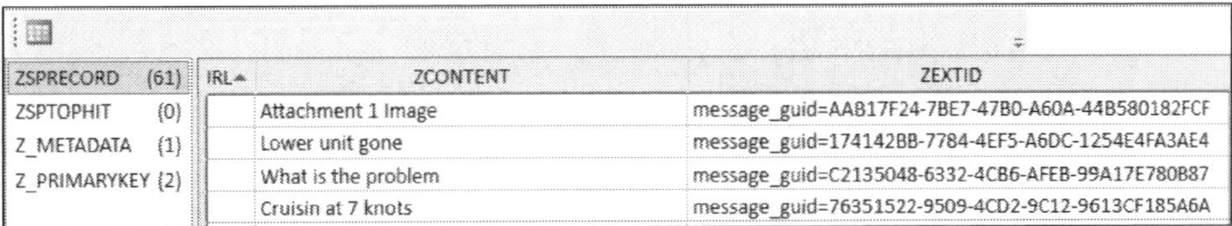

IRL	ZCONTENT	ZEXTID
	Attachment 1 Image	message_guid=AAB17F24-7BE7-47B0-A60A-44B580182FCF
	Lower unit gone	message_guid=174142BB-7784-4EF5-A6DC-1254E4FA3AE4
	What is the problem	message_guid=C2135048-6332-4CB6-AFEB-99A17E780B87
	Cruisin at 7 knots	message_guid=76351522-9509-4CD2-9C12-9613CF185A6A

The SMS Spotlight Cache file

You can run the following commands to dump the SMS Spotlight cache database into a CSV file named `smsspotlightcache.csv`:

```
$sqlite3 smssearchindex.sqlite
SQLite version 3.7.12 2012-04-03 19:43:07
Enter ".help" for instructions
Enter SQL statements terminated with a ";"
sqlite>.mode csv
sqlite>.output smsspotlightcache.csv
sqlite>.headers on
sqlite> SELECT * FROM Content;
sqlite>.exit
```

Calendar events

Calendar events that have been manually created by the user or synced using a mail application or other third-party applications are stored in the calendar database. The calendar database is a `HomeDomain` file and can be found at `/private/var/mobile/Library/Calendar/Calendar.sqlitedb`.

The `CalendarItem` table in the `Calendar.sqlitedb` file contains the calendar events summary, description, start date, end date, and more. You can run the following command lines to dump the calendar database into a CSV file named `calendar.csv`. Note that reminders and tasks are often saved in the `Calendar.sqlitedb` file. These files may not contain a start or end time depending on the event:

```
$ sqlite3 Calendar.sqlitedb
SQLite version 3.7.12 2012-04-03 19:43:07
Enter ".help" for instructions
Enter SQL statements terminated with a ";"
sqlite>.mode csv
sqlite>.output calendar.csv
sqlite>.headers on
```

```
sqlite> SELECT
            rowid,summary,description,
            datetime(start_date + 978307200,'unixepoch') as
            start_date,
            datetime (end_date + 978307200,'unixepoch') as
end_date
        FROM CalendarItem;
sqlite>.exit
```

E-mail database

All e-mail or mail applications on the device are stored in a SQLite database file. The database is a `HomeDomain` file and can be found at `/private/var/mobile/Library/Mail/Protected Index`. The database file has no extension and contains locally stored, sent, and deleted messages.

You can run the following commands to obtain e-mails stored in the mail database:

```
$ sqlite3 Protected\ Index
SQLite version 3.7.12 2012-04-03 19:43:07
Enter ".help" for instructions
Enter SQL statements terminated with a ";"
sqlite>.output Email.csv
sqlite>.headers on
sqlite> SELECT * FROM messages;
sqlite>.exit
```

In addition to the messages, e-mail attachments are also often stored on the file system within the `Mail` directory.

Notes

The `Notes` database contains the notes created by the user using the device's built-in **Notes** application. Notes is the simplest application, often containing the most sensitive and confidential information. The `Notes` database is a `HomeDomain` file and can be found at `/private/var/mobile/Library/Notes/notes.sqlite`.

The `Znote` and `Znotebody` tables in the `notes.sqlite` file contain the notes title, content, creation date, modification date, and more. You can run the following commands to dump the `Notes` database into a CSV file named `notes.csv`:

```
$sqlite3 notes.sqlite
SQLite version 3.7.12 2012-04-03 19:43:07
```

```
Enter ".help" for instructions
Enter SQL statements terminated with a ";"
sqlite>.mode csv
sqlite>.output notes.csv
sqlite>.headers on
sqlite> SELECT
         datetime(zcreationdate+978307200,'unixepoch') as
         zcreationdate,
         datetime(zmodificationdate+978307200,'unixepoch') as
         zmodificationdate,
         ztitle, zsummary, zcontent
       FROM znote, znotebody
       WHERE znotebody.z_pk=znote.z_pk
       ORDER BY znote.z_pk ASC;
sqlite>.exit
```

Safari bookmarks

The **Safari** browser used on an Apple device allows users to bookmark their favorite websites. The bookmarks database is a `HomeDomain` file and can be found at `/private/var/mobile/Library/Safari/Bookmarks.db`.

You can run the following commands to view the bookmarks stored in the database:

```
$sqlite3 bookmarks.db
SQLite version 3.7.12 2012-04-03 19:43:07
Enter ".help" for instructions
Enter SQL statements terminated with a ";"
sqlite>.headers on
sqlite> select title, url from bookmarks;
sqlite>.exit
```

The Safari web caches

The Safari browser stores the recently downloaded and cached data in a database. The database is a `HomeDomain` file and can be found at `/private/var/mobile/Library/Caches/com.apple.mobilesafari/Cache.db`. The file contains cached URLs and the web server's responses along with the timestamps.

The web application cache

Offline data cached by web applications, such as images, HTML, JavaScript, style sheets, and more are stored in a database. The database is a `HomeDomain` file and can be found at `/private/var/mobile/Library/Caches/com.apple.WebAppCache/ApplicationCache.db`.

The WebKit storage

Safari stores information from various sites in the `WebKit` database located in the `/private/var/mobile/Library/WebKit/LocalStorage/` directory. The directory contains unique databases for each website, as shown in the following screenshot:

```
mbp-hmahalik:Webkit hmahalik$ cd /Users/
hmahalik/Desktop/Webkit/LocalStorage
mbp-hmahalik:LocalStorage hmahalik$ ls
StorageTracker.db
http_www.google.com_0.localstorage
http_m.youtube.com_0.localstorage
http_www.youtube.com_0.localstorage
http_www.bing.com_0.localstorage
https_m.facebook.com_0.localstorage
mbp-hmahalik:LocalStorage hmahalik$
```

The LocalStorage folder contents

The photos metadata

A manifestation of the photos in the device's photo album is stored in a database located at `/private/var/mobile/Media/PhotoData/Photos.sqlite`. The photos metadata database file is a member of `CameraRollDomain`.

You can run the following commands to view the photos stored in the database:

```
$sqlite3 Photos.sqlite
SQLite version 3.7.12 2012-04-03 19:43:07
Enter ".help" for instructions
Enter SQL statements terminated with a ";"
sqlite>.mode csv
sqlite>.output photos.csv
sqlite>.headers on
sqlite> SELECT
            z_pk, ztitle,
            datetime(zdatecreated + 978307200,'unixepoch') as
            zdatecreated,
            datetime(zmodificationdate+ 978307200,'unixepoch') as
            zmodificationdate,
            zfilename, zdirectory, zwidth, zheight
```

```
        FROM zgenericasset
        ORDER BY z_pk ASC;
sqlite>.exit
```

Consolidated GPS cache

Geolocation history of cell towers and Wi-Fi on the device is stored in one of the two possible databases that are located at `/private/var/root/Caches/locationd/`. The databases are either `consolidated.db` or `cache_encryptedA.db`. Both database files are members of `RootDomain`. The version of iOS will determine which database is used. These databases contain location information for cell towers that the device came into close proximity with as well as Wi-Fi networks that were available for the device to connect to. These databases are often used to place a person near a specific location as this data is cached to one of these database files without the user's consent.

For this example, we will examine the `consolidated.db` file. The `CompassCalibration` table in the `consolidated.db` file contains the location information along with the timestamps. The file, when opened with SQLite Professional, displays the data as shown in the following screenshot. Note that the `cache_encryptedA.db` file is no longer backed up when the user syncs with iTunes.

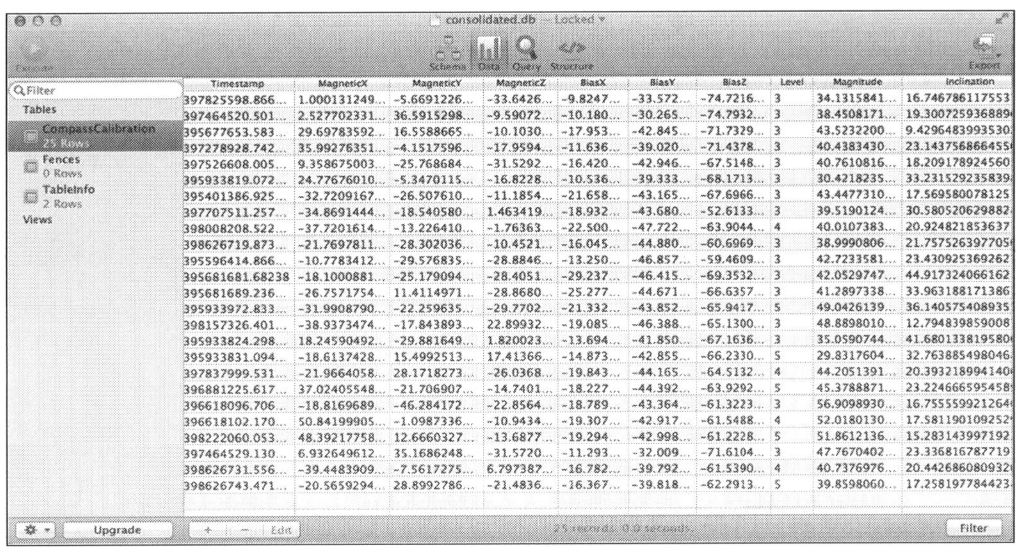

Timestamp	MagneticX	MagneticY	MagneticZ	BiasX	BiasY	BiasZ	Level	Magnitude	Inclination
397825598.866...	1.000131249...	-5.6691226...	-33.6426...	-9.8247...	-33.572...	-74.7216...	3	34.1315841...	16.746786117553
397464520.501...	2.527702331...	36.5915298...	-9.59072...	-10.180...	-30.265...	-74.7932...	3	38.4508171...	19.300725936889
395677653.583...	29.69783592...	16.5588665...	-10.1030...	-17.953...	-42.845...	-71.7329...	3	43.5232200...	9.4296483993530
397278928.742...	35.99276351...	-4.1517596...	-17.9594...	-11.636...	-39.020...	-71.4378...	3	40.4383430...	23.143756866455
397526608.005...	9.358675003...	-25.768684...	-31.5292...	-16.420...	-42.946...	-67.5148...	3	40.7610816...	18.209178924560
395933819.072...	24.77676010...	-5.3470115...	-16.8228...	-10.536...	-39.333...	-68.1713...	3	30.4218235...	33.231529235839
395401386.925...	-32.7209167...	-26.507610...	-11.1854...	-21.658...	-43.165...	-67.6966...	3	43.4477310...	17.569580078125
397707511.257...	-34.8691444...	-18.540580...	1.463419...	-18.932...	-43.680...	-52.6133...	3	39.5190124...	30.580520629882
398008208.522...	-37.7201614...	-13.226410...	-1.76363...	-22.500...	-47.722...	-63.9044...	4	40.0107383...	20.924821853637
398626719.873...	-21.7697811...	-28.302036...	-10.4521...	-16.045...	-44.880...	-60.6969...	3	38.9990806...	21.757526397705
395596414.866...	-10.7783412...	-29.576835...	-28.8846...	-13.250...	-46.857...	-59.4609...	3	42.7233581...	23.430925369262
395681681.68238	-18.1000881...	-25.179094...	-28.4051...	-29.237...	-46.415...	-69.3532...	3	42.0529747...	44.917324066162
395681689.236...	-26.7571754...	11.4114971...	-28.8680...	-25.277...	-44.671...	-66.6357...	3	41.2897338...	33.963188171386
395933972.833...	-31.9908790...	-22.259635...	-29.7702...	-21.332...	-43.852...	-65.9417...	5	49.0426139...	36.140575408935
398157326.401...	-38.9373474...	-17.843893...	22.89932...	-19.085...	-46.388...	-65.1300...	3	48.8898010...	12.794839859008
395933824.298...	18.24590492...	-29.881649...	1.820023...	-13.694...	-41.850...	-67.1636...	3	35.0590744...	41.680133819580
395933831.094...	-18.6137428...	15.4992513...	17.41366...	-14.873...	-42.855...	-66.2330...	5	29.8317604...	32.763885498046
397837999.531...	-21.9664058...	28.1718273...	-26.0368...	-19.843...	-44.165...	-64.5132...	4	44.2051391...	20.393218994140
396881225.617...	37.02405548...	-21.706907...	-14.7401...	-18.227...	-44.392...	-63.9292...	5	45.3788871...	23.224666595458
396618096.706...	-18.8169689...	-46.284172...	-22.8564...	-18.789...	-43.364...	-61.3223...	3	56.9098930...	16.755559921264
396618102.170...	50.84199905...	-1.0987336...	-10.9434...	-19.307...	-42.917...	-61.5488...	4	52.0180130...	17.581190109252
398222060.053...	48.39217758...	12.6660327...	-13.6877...	-19.294...	-42.998...	-61.2228...	5	51.8612136...	15.283143997192
397464529.130...	6.932649612...	35.1686248...	-31.5720...	-11.293...	-32.009...	-71.6104...	3	47.7670402...	23.336816787719
398626731.556...	-39.4483909...	-7.5617275...	6.797387...	-16.782...	-39.792...	-61.5390...	4	40.7376976...	20.442686080932
398626743.471...	-20.5659294...	28.8992786...	-21.4836...	-16.367...	-39.818...	-62.2913...	5	39.8598060...	17.258197784423

The Consolidated.db view with SQLite Professional

Voicemail

The `voicemail` database contains metadata about each voicemail stored on the device that includes the sender's phone number, callback number, timestamp and message duration, and more. The voicemail recordings are stored as `AMR` audio files that can be played by any media player that supports the AMR codec (for example, **QuickTime Player**). The `voicemail` database is a `HomeDomain` file and can be found at `/private/var/mobile/Library/Voicemail/voicemail.db`, while the actual voicemail recordings are stored in the `/private/var/mobile/Library/Voicemail/` directory.

You can run the following commands to view the list of voicemails stored in the database:

```
$sqlite3 voicemail.sqlite
SQLite version 3.7.12 2012-04-03 19:43:07
Enter ".help" for instructions
Enter SQL statements terminated with a ";"
sqlite>.headers on
sqlite> SELECT * FROM voicemail;
sqlite> .exit
```

Property lists

A property list, commonly referred to as a plist, is a structured data format used to store, organize, and access various data types of data on an iOS device as well as a Mac OS X device. Plists are binary-formatted files and can be viewed using a **Property List Editor**, which is capable of reading or converting the binary format to ASCII.

Plist files may or may not have a `.plist` file extension. To access the data stored in these files, you need a tool that can read them. Some of the good free tools include:

- Plist Editor for Windows, which can be downloaded from `http://www.icopybot.com/plist-editor.htm`
- The plutil command-line utility on Mac OS X

You can also view the plist files using XCode. Mac OS X includes the plutil command-line utility by default. The command-line utility can easily convert the binary formatted files into human readable files.

The following example displays the Safari browser `History.plist` file:

```
$sudo plutil -convert xml1 History.plist -o -
<?xml version="1.0" encoding="UTF-8"?>
<!DOCTYPE plist PUBLIC "-//Apple//DTD PLIST 1.0//EN"
"http://www.apple.com/DTDs/PropertyList-1.0.dtd">
<plist version="1.0">
<dict>
  <key>WebHistoryDates</key>
  <array>
    <dict>
      <key></key>
      <string>http://www.securitylearn.net/</string>
      <key>D</key>
      <array>
        <integer>1</integer>
      </array>
      <key>lastVisitedDate</key>
      <string>411576251.8</string>
      <key>title</key>
      <string>securitylearn/</string>
      <key>visitCount</key>
      <integer>1</integer>
    </dict>
    <dict>
      <key></key>
      <string>http://www.google.com</string>
      <key>D</key>
      <array>
      <integer>1</integer>
      </array>
      <key>lastVisitedDate</key>
      <string>411571510.5</string>
      <key>title</key>
      <string>Google</string>
      <key>visitCount</key>
      <integer>1</integer>
    </dict>
  </array>
```

```
  <key>WebHistoryFileVersion</key>
  <integer>1</integer>
</dict>
</plist>
```

Important plist files

Raw disk images or the backup that you extracted in *Chapter 3, Data Acquisition from iOS Devices*, and *Chapter 4, Data Acquisition from iOS Backups*, will contain the following plist files that are important for an investigation. The files shown are extracted from an iOS 6 device. The file locations may vary for your iOS version.

The HomeDomain plist files

The following are the `HomeDomain` plist files, which contain data that may be relevant to your investigation:

- `/private/var/mobile/Library/Preferences/com.apple.mobilephone.plist`: This contains the last phone number entered into the dialer regardless of whether it was dialed or not
- `/private/var/mobile/Library/Preferences/com.apple.mobilephone.speeddial.plist`: This contains a list of the contacts that were added to the phone's favorite list
- `/private/var/mobile/Library/Preferences/com.apple.accountsettings.plist`: This contains a list of the e-mail accounts configured on the device
- `/private/var/mobile/Library/Preferences/com.apple.AppSupport.plist`: This contains the country code used for the App Store on the device
- `/private/var/mobile/Library/Preferences/com.apple.Maps.plist`: This contains the last latitude, longitude, and address pinned in the Maps application
- `/private/var/mobile/Library/Preferences/com.apple.mobilemail.plist`: This contains the e-mail fetching dates and e-mail signatures used
- `/private/var/mobile/Library/Preferences/com.apple.mobiletimer.plist`: This contains a list of world clocks used
- `/private/var/mobile/Library/Preferences/com.apple.Preferences.plist`: This contains the keyboard language that was last used on the device
- `/private/var/mobile/Library/Preferences/com.apple.mobilesafari.plist`: This contains a list of the recent searches made through Safari

- /private/var/mobile/Library/Preferences/Com.apple.springboard.plist:This contains a list of applications that are shown in the interface and iOS version
- /private/var/mobile/Library/Preferences/com.apple.mobiletimer.plist: This contains information about the current time zone, timers, alarms, and stopwatches
- /private/var/mobile/Library/Preferences/com.apple.weather.plist: This contains the cities for weather reports, date, and time of last update
- /private/var/mobile/Library/Preferences/com.apple.stocks.plist: This contains a list of the stocks tracked
- /private/var/mobile/Library/Preferences/com.apple.preferences.network.plist: This contains the status of Bluetooth and Wi-Fi networks
- /private/var/mobile/Library/Preferences/com.apple.conference.history.plist: This contains a history of the phone numbers and other accounts that were conferenced using FaceTime
- /private/var/mobile/Library/Preferences/com.apple.locationd.plist: This contains a list of application identifiers that use the location service on the device
- /private/var/mobile/Library/Safari/History.plist: This contains the web browsing history of Safari
- /private/var/mobile/Library/Safari/SuspendState.plist: This contains the web page title and the URL of all suspended web pages on Safari
- /private/var/mobile/Library/Maps/Bookmarks.plist: This contains the bookmarked locations within the Maps application
- /private/var/mobile/Library/Caches/com.apple.mobile.installation.plist: This contains a list of all system and user applications loaded onto the device and their disk paths
- /private/var/mobile/Library/Caches/com.apple.UIKit.pboard/pasteboard: This contains a cached copy of the data stored on the device's clipboard

The RootDomain plist files

The following RootDomain files listed should be examined for relevance to your investigation:

- /private/var/root/Library/Preferences/com.apple.preferences.network.plist: This contains information about whether the airplane mode is presently enabled on the device

- `/private/var/root/Library/Lockdown/pair_records`: This directory contains property lists with private keys used in order to pair the device to a computer
- `/private/var/root/Library/Caches/locationd/clients.plist`: This contains the location settings for applications and system services

The WirelessDomain plist files

The following `WirelessDomain` plist file contains useful information to identify the SIM card last used in the device:

- `/private/wireless/Library/Preferences/com.apple.commcenter.plist`

The SystemPreferencesDomain plist files

The two plist files containing data of evidentiary value from the `SystemPreferencesDomain` files are listed:

- `/private/var/preferences/SystemConfiguration/com.apple.network.identification.plist`: This contains networking information of the cached IP
- `/private/var/preferences/SystemConfiguration/com.apple.wifi.plist`: This contains a list of previously known Wi-Fi networks and the last time each one was connected to

Other important files

Apart from SQLite and plist files, several other locations may contain valuable information to an investigation.

The others sources include the following:

- Cookies
- Keyboard cache
- Photos
- Wallpaper
- Snapshots

- Recordings
- Downloaded applications

Cookies

Cookies can be recovered from `/private/var/mobile/Library/Cookies/Cookies.binarycookies`. This file is a standard binary file that contains cookies saved when web pages are accessed on the device. This information can be a good indication of what websites the user has been actively visiting.

To convert the binary cookie to human readable format, run the `BinaryCookieReader.py` Python script on the cookie file, as shown in the following command. The Python script source code is available in the code bundle of the book.

```
$python BinaryCookieReader.py Cookies.binarycookies
```

```
Cookie :
__utma=167051323.813879307.1359034257.1367989551.1386632713.9;
domain=.testflightapp.com; path=/; expires=Wed, 09 Dec 2015;

Cookie : __utmb=167051323.24.8.1386633092975;
domain=.testflightapp.com; path=/; expires=Tue, 10 Dec 2013;

Cookie :
__utmz=167051323.1386632713.9.1.utmcsr=(direct)|utmccn=(direct)|utmcm
d=(none); domain=.testflightapp.com; path=/; expires=Tue, 10 Jun
2014;

Cookie : tfapp=1d29da4a798a90186f1d4bfce3ce2f23;
domain=.testflightapp.com; path=/; expires=Thu, 09 Feb 2017;

Cookie : user_segment=Prospect; domain=.testflightapp.com; path=/;
expires=Wed, 08 Jan 2014;
[...]
```

Keyboard cache

Keyboard cache is captured and saved in the `dynamic-text.dat` file. The file is located at `/private/var/mobile/Library/Keyboard/dynamic-text.dat` and contains keyboard cache, which comprises of text entered by the user. This text is cached as part of the device's autocorrect feature and was designed to autocomplete the predictive common words. The file keeps a list of approximately 600 words per language used on the iOS device.

It is a binary file and can be viewed using a hex editor, as shown in the following screenshot. This file may contain passwords cached by the iOS device and can be used to achieve brute force attacks on the device or an encrypted backup of the device.

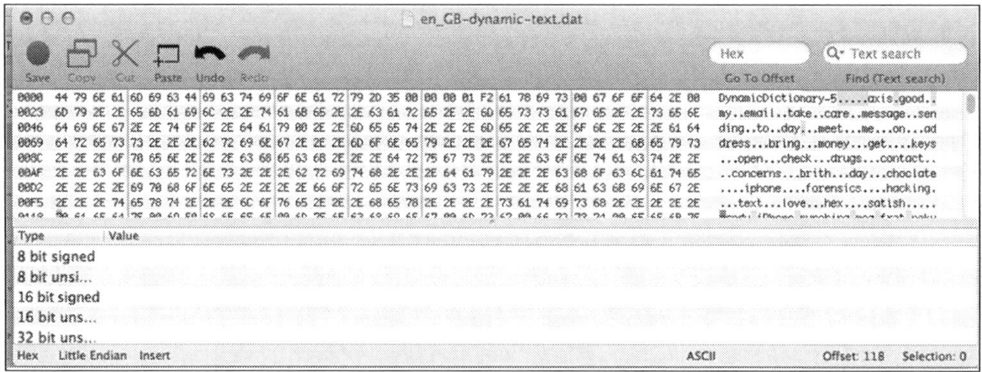

Keyboard cache in hex editor

Photos

Photos are stored in a directory located at `/private/var/mobile/Media/DCIM/`, which contains the photos taken with the device's built-in camera, screenshots, and accompanying thumbnails. Some third-party applications will also store photos taken in this directory. Every photo stored in the `DCIM` folder contains `EXIF` (Exchangeable Image File Format) data. EXIF data stored in the photo can be extracted using **exiftool**, which can be downloaded from `http://www.sno.phy.queensu.ca/~phil/exiftool/`. EXIF data may also contain the geographical information when a photo is tagged with the user's geo location if the user has enabled location permissions on the iOS device:

```
$exiftool IMG_0107.JPG
ExifTool Version Number         : 9.50
File Name                       : IMG_0107.JPG
Directory                       : .
File Size                       : 73 kB
File Modification Date/Time     : 2014:01:07 17:43:05+05:30
File Access Date/Time           : 2014:02:09 17:26:40+05:30
File Inode Change Date/Time     : 2014:02:09 17:26:40+05:30
File Permissions                : rw-r--r--
[...]
```

Wallpaper

The current background wallpaper set for the iOS device can be recovered from the `LockBackgroundThumbnail.jpg` file found in `/private/var/mobile/Library/SpringBoard/LockBackground.cpbitmap`. This is complemented with a thumbnail named in the same directory. The wallpaper picture may contain identifying information about the user, which could help in a missing person's case or an iOS device recovered from a theft investigation.

Snapshots

The `snapshots` directory contains screenshots of the most recent states of built-in applications at the time that they were suspended. This directory is located in `/private/var/mobile/Library/Caches/Snapshots/`. Every time an application is suspended to the background by pressing the **Home** button, a snapshot is taken to produce a nice shrinking effect. Third-party applications also store the snapshot cache inside their application's folder.

Recordings

The iPhone allows a user to record voice memos very easily. The recorded voice memos are stored in the `/private/var/mobile/Media/Recordings/` directory. Recordings here could be used to identify a person based upon their voice and may also contain information such as voice reminders, which won't be stored in the calendar database. Recordings provide a lot of information to the examiner as they are user created and often not deleted.

Downloaded applications

Third-party applications, which are downloaded and installed from the App Store, include applications such as Facebook, WhatsApp, Viber, Wickr, Skype, and GMail, and more that contain a wealth of information useful for an investigation. Some third-party applications use the `Base64` encoding, which needs to be converted for viewing as well as encryption. Applications that encrypt the database file prevent the examiner from accessing the data residing in the tables. Encryption varies amongst these applications based on the application and iOS versions.

A unique subdirectory GUI is created for each application installed on the device in the `/private/var/mobile/Applications/` directory, which is shown in the following example. Also, the hierarchical structure of the `Applications` directory is shown. Most of the files stored in the application's directory are in the SQLite and plist format:

```
$tree -L 2 /var/mobile/Applications/
/var/mobile/Applications/
|-- 08E03CB2-26A5-4DAF-9843-3893AF4EDDF0
|    |-- Documents
|    |-- Library
|    |-- WordPress.app
|    |-- iTunesArtwork|    |-- iTunesMetadata.plist
|    `-- tmp
|-- 0922F95C-7E40-4075-BC5A-06CE829BDD9E
|    |-- Documents
|    |-- Library
|    |-- Wickr.app
|    |-- iTunesArtwork
|    |-- iTunesMetadata.plist
|    `-- tmp
|-- 11C7F3E9-A10E-405D-B6BB-2F86B1B2400F
|    |-- Documents
|    |-- Library
|    |-- photovault.app
|    `-- tmp
```

Recovering deleted SQLite records

In addition to the recovering techniques covered in *Chapter 3, Data Acquisition from iOS Devices*, you can also recover the deleted records from a SQLite database. SQLite databases store the deleted records within the database itself. So, it is possible to recover the deleted data such as contacts, SMS, calendar, notes, e-mails and voicemails, and more by parsing the corresponding SQLite database. If a SQLite database is vacuumed or defragmented, the likelihood of recovering the deleted data is minimal. The amount of cleanup these databases require heavily relies on the iOS version, the device, and the user's settings on the device.

A SQLite database file comprises one or more fixed size pages, which are used just once. SQLite uses a **b-tree** layout of pages to store indices and table content. Detailed information on the b-tree layout is explained at `http://sandbox.dfrws.org/2011/fox-it/DFRWS2011_results/Report/Sqlite_carving_extractAndroidData.pdf`.

To carve a SQLite database, you can examine the data in raw hex or use `sqliteparse.py`, a Python script developed by Mari DeGrazia. The Python script can be downloaded from `http://www.arizona4n6.com/download/SQLite-Parser.zip`.

The following example recovers the deleted records from the `notes.sqlitedb` file and dumps the output to the `output.txt` file. To validate your findings from running the script, simply examine the database in a hex viewer to ensure nothing is overlooked:

```
$python sqliteparse.py -f notes.sqlitedb -r -o output.txt
```

In addition to it, performing a `strings` dump of the database file can also reveal deleted records that may have been missed, as shown in the following command:

```
$strings notes.sqlitedb
```

Summary

This chapter covered various data analysis techniques and specified the locations of data within the iOS device's file system. We also explained most of the common file formats used in the iPhone and walked you through important files to recover the most data possible. Most open source and commercial tools are able to pull deleted data from common database files, such as contacts, calls, SMS, and more, but they often overlook the third-party application database files. We covered techniques to recover deleted SQLite records that prove useful in most iOS device investigations. Again, the acquisition method, encoding, and encryption schemas can affect the amount of data you can recover during your examination. In the next chapter, we will discuss iOS forensic tools, which will help you acquire and analyze data.

6

iOS Forensic Tools

Although understanding acquisition methods and techniques is helpful, a forensic examiner often needs the help of tools to accomplish tasks in the given time. Forensic tools not only save time but also make the process a lot easier. Currently, there are many commercial tools such as Elcomsoft iOS Forensic Toolkit, Cellebrite UFED, BlackLight, Oxygen Forensic Suite, AccessData MPE+, iXAM, Lantern, XRY, SecureView, Paraben iRecovery Stick, and so on, which are available for forensic acquisition and analysis of an iOS device. For familiarity purposes, this chapter will walk you through the usage of a few commercial and open source tools and provide details of the steps required to perform acquisitions of iOS devices.

Elcomsoft iOS Forensic Toolkit

Elcomsoft iOS Forensic Toolkit (**EIFT**) is a set of tools aimed at making the acquisition of iOS devices easier. EIFT is a combination of software that is able to perform forensic acquisition of iOS devices running any version of iOS (note: some iOS versions require the device to be jailbroken). EIFT can acquire bit-for-bit images of a device's file system, extract device secrets (passcodes, passwords, and encryption keys), and decrypt the file system image. For more information on EIFT, visit `http://www.elcomsoft.com/eift.html`.

The toolkit was initially available only to law enforcement agencies, but now it is available to everyone. The toolkit supports both Mac OS X and Windows platforms with iTunes 10.6 or later installed.

Features of EIFT

The following are the features of EIFT:

- Supports physical and logical acquisition.
- Acquires complete bit-for-bit device images.

- Quick file system acquisition: 20-40 minutes for 32 GB models.
- Supports passcode recovery attacks.
- Extracts device keys required to decrypt a raw disk image as well as keychain items.
- Decrypts a raw disk image and keychain items.
- Zero-footprint: this operation leaves no traces and alterations to device contents.
- Fully accountable: every step of investigation is logged and recorded.

Usage of EIFT

Elcomsoft iOS Forensic Toolkit can be used in two modes: guided mode and manual mode. The USB dongle shipped with the toolkit must be connected to the computer while the toolkit is running.

Guided mode

The guided mode features a menu-based user interface where you can accomplish typical tasks by selecting the corresponding menu items. You can start the guided mode by double-clicking on the `Toolkit.cmd` (Windows) or `Toolkit.command` (Mac OS X) file in the directory where you have copied the toolkit files. This should open the terminal window and present a text-based menu as shown in the following screenshot:

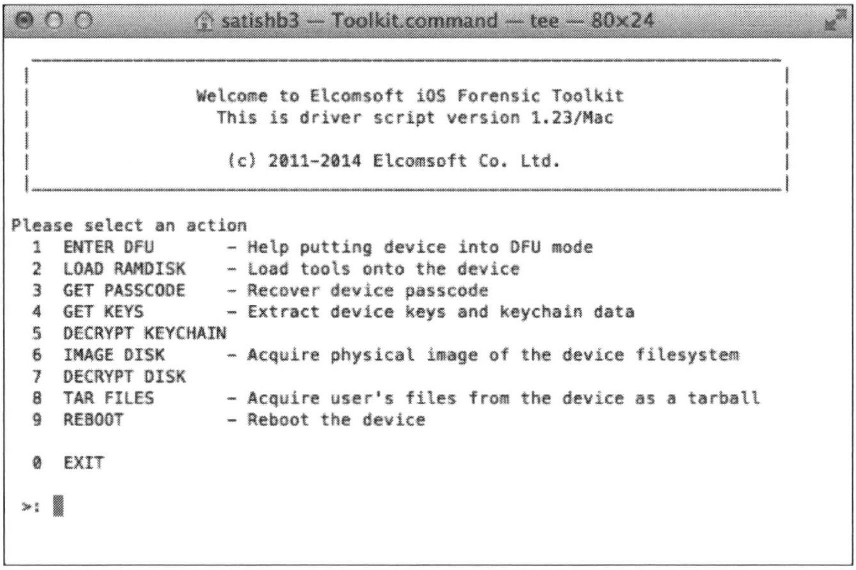

```
satishb3 — Toolkit.command — tee — 80x24

 _____________________________________________________________________
|                                                                     |
|               Welcome to Elcomsoft iOS Forensic Toolkit             |
|                 This is driver script version 1.23/Mac              |
|                                                                     |
|                  (c) 2011-2014 Elcomsoft Co. Ltd.                   |
|_____________________________________________________________________|

Please select an action
  1  ENTER DFU        - Help putting device into DFU mode
  2  LOAD RAMDISK     - Load tools onto the device
  3  GET PASSCODE     - Recover device passcode
  4  GET KEYS         - Extract device keys and keychain data
  5  DECRYPT KEYCHAIN
  6  IMAGE DISK       - Acquire physical image of the device filesystem
  7  DECRYPT DISK
  8  TAR FILES        - Acquire user's files from the device as a tarball
  9  REBOOT           - Reboot the device

  0  EXIT

>:
```

The Elcomsoft iOS Forensic Toolkit welcome screen

When running in the guided mode, the toolkit logs all the activities to a text file. Each time the toolkit is started, a new log file is created in the user's home directory and the output of all the invoked commands as well as user choices are written to that file.

To perform the physical acquisition of iPhone 4 and older devices with EIFT, follow the steps provided:

1. Put the device in the DFU mode. You can do this by selecting the menu item **1** and following the onscreen instructions.
2. After the device has been put in the DFU mode, load the ramdisk with the acquisition tools by selecting menu item **2** or answer `y` to the prompt that follows the DFU procedure. It automatically detects the type of the device and loads the compatible ramdisk onto it. When ramdisk is successfully loaded, the device screen will show the Elcomsoft logo.
3. Recover the device passcode by selecting menu item **3**. The toolkit can recover a simple 4-digit passcode in less than 20 minutes. It also provides options to perform dictionary (wordlist) and brute force attacks on complex passwords, as shown in the following screenshot:

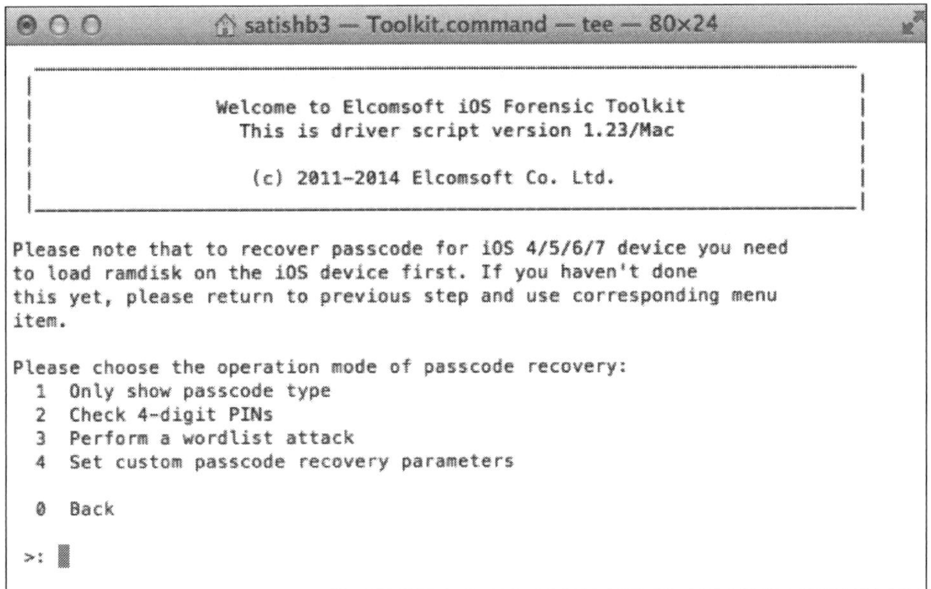

The EIFT passcode recovery options

4. Extract the encryption keys required to decrypt files and keychain items by selecting menu item **4**. You will be prompted to supply the device passcode, if known, of the escrow file if you have access to the host computer and a filename to save the keys. If the filename is not supplied, the toolkit extracts the keys and stores it in the `keys.plist` file in the user's home directory.
5. After extracting the keys, to decrypt the keychain items, select menu item **5**. The toolkit uses the keys stored in the `keys.plist` file, decrypts the keychain items, and stores it in the `keychain.txt` file in the user's home directory.
6. To acquire the physical image of the device's file system, select the menu item **6**. You will be prompted to choose the device partition (system and user data) to image, as shown in the following screenshot:

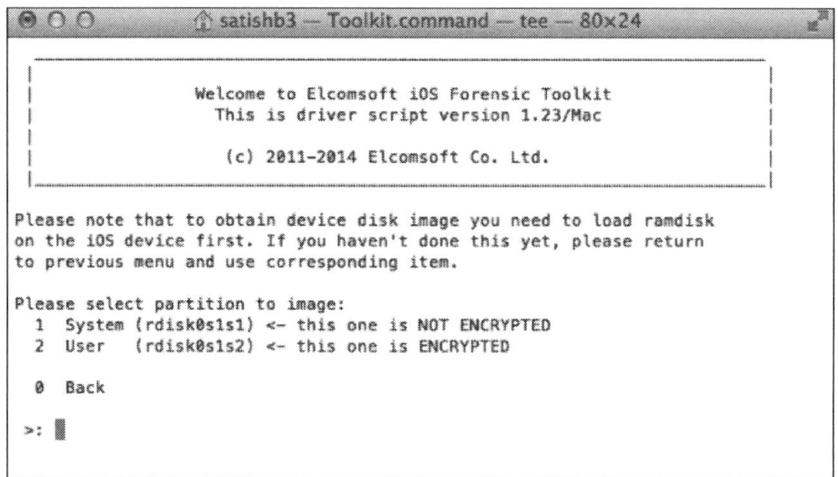

EIFT – selecting partition to image option

After selecting the partition, the window prompts you for a filename to save the image. If the filename is not supplied, it extracts the raw file system from the device and stores it as a `user.dmg` file in the user's home directory. Best practices include acquiring both the user and system partitions.

7. After the acquisition, you can reboot the device to function normally by selecting menu item **9**.
8. To decrypt the acquired image, select menu item **7**. You will be prompted to provide filenames of the encrypted image, device keys, and a filename to save the decrypted image. If the filename is not supplied, it decrypts the image and stores it as `user-decrypted.dmg` in the user's home directory. The toolkit also computes the SHA1 hash of the decrypted image file. EIFT is also capable of performing physical acquisition of a jailbroken iPhone 4S and newer devices running on iOS 5/6/7. At the time of writing this, EIFT is the only tool that supports physical acquisition of the iPhone 4S and newer devices running with iOS 7. EIFT requires the OpenSSH package to be installed on the device to perform acquisition on newer devices. OpenSSH runs the SSH server on the device and allows you to copy and run the acquisition tools. Once the SSH server is running on the device, you can follow steps 3 to 8 to acquire a raw disk image from an iPhone 4S and newer devices.

Manual mode

The manual mode lets you interact with tools directly using the command-line interface. This mode allows greater flexibility and is recommended if you are comfortable with using command-line tools. The commands required to accomplish typical tasks in the manual mode are well documented in the technical guide that comes with the toolkit.

The toolkit is capable of performing physical and logical acquisition of the device's file system. But it does not provide options to analyze the acquired data and recover the deleted data. However, you can supply the `.dmg` file acquired with EIFT to Oxygen Forensic Suite, Cellebrite Physical Analyzer, and other tools for data analysis and recovery.

EIFT-supported devices

Elcomsoft iOS Forensic Toolkit Version 1.23 supports most iOS devices, however some must be jailbroken. The following figure is taken directly from the help document that comes with the toolkit:

		Physical imaging	Logical imaging	Passcode recovery	Keychain decryption	Disk decryption
iPhone iPhone 3G iPod Touch 1 iPod Touch 2	iOS 1/2/3	+	+	instant 2)	+	not encrypted 3)
	iOS 4	+	+	+	+	not encrypted 3)
iPhone 3GS iPod Touch 3 iPad 1	iOS 3	+	+	instant 2)	+	not encrypted 3)
	iOS 4/5	+	+	+	+	+ 4)
iPhone 4 iPod Touch 4	iOS 4/5/6/7	+	+	+	+	+
iPhone 4S iPhone 5 iPhone 5C iPad 2-4 iPad Mini iPod Touch 5	iOS 5/6/7	+	+	+	+	+

EIFT supported devices

Compatibility notes

The following are the compatibilities of EIFT-supported devices:

- Support for iPhone 4S/5/5S/5C, iPad 2 and later versions, and iPod Touch 5th generation devices is currently limited to jailbroken devices.
- iOS versions before 3.x store the device passcode in the keychain. On these devices, the passcode is recovered instantly during the encryption key and keychain data recovery.
- Devices running iOS versions before 3.x do not have data protection enabled and user partition is not encrypted.

- If a device was shipped with iOS 3.x installed and was updated to iOS 4.x without reset (which erases all contents and settings), that is, using the **Update** option in iTunes instead of **Restore**, then data protection is not enabled and the user partition is not encrypted.

Oxygen Forensic Suite 2014

Oxygen Forensic Suite 2014 is an advanced forensic software to extract and analyze data from cell phones, smartphones, PDAs, and other mobile devices. The software provides logical support for the widest range of mobile devices and allows fully automated forensic acquisition and analysis. Currently, Oxygen Forensic Suite 2014 Version 6.1 supports more than 7,700 different model mobile devices.

Oxygen Forensic Suite 2014 uses proprietary low-level protocols to extract data from smartphones. Besides data extraction, Oxygen Forensic Suite also gives you the opportunity to import a backup/image file obtained using other forensic tools, such as Cellebrite, Elcomsoft, XRY, iTunes, and Lantern Lite for data analysis. It also stores the database of all the analyzed devices, so you can always view the previously extracted data or use a powerful multiphone search feature to find the required details.

Oxygen Forensic Suite 2014 is available only for the Windows platform and requires iTunes to be installed on the computer. The software costs $2,999 for the full version, and a freeware version is also available with limited functionalities. The software operates with original and jailbroken devices and extracts the following data: phonebook with assigned photos, calendar events and notes, call logs, messages, camera snapshots, video and music, voice mail, passwords, dictionaries, geopositioning data, Wi-Fi points with passwords and coordinates, IP connections, locations, navigation applications, device data, factory installed, third-party applications data, and so on. It also recovers deleted data from SQLite databases and can recover calls, messages, e-mail messages, e-mail accounts, photo thumbnails, contact photos, and so on. This tool does not support physical acquisition, thus a full forensic image cannot be obtained. For more information, visit `http://www.oxygen-forensic.com/de/compare/devices/software-for-iphone`.

Features of Oxygen Forensic Suite

The following are the features of Oxygen Forensic Suite:

- It supports logical acquisition. Logical acquisition recovers the active files on the device. Deleted data may be obtained if the SQLite database is recovered. Physical and file system acquisition are not supported by this tool. Both of these acquisition methods provide access to the raw file system data of the iOS device.
- Password recovery from a keychain.
- Read backup/images obtained using other forensic tools.
- Timeline: This provides a single-place access to all the user's activities and movements arranged by date and time.
- Zero-footprint operation: This leaves no traces and alterations to device contents.
- It supports aggregated contacts. This automatically combines accounts from different sources in one metacontact for each person. (Caution: Make sure you know where the data is coming from! You should manually examine each file to ensure nothing is overlooked and that the data is being reported correctly.)
- It recovers deleted data automatically.
- It provides access to raw files for manual analysis. (Note: These are the raw database files associated with each application, not the raw file system partitions.)
- It provides an intuitive and user-friendly UI to browse the extracted data.
- It provides keyword lists and a regular expression library in order to search.
- Report generation in several popular formats—Microsoft Excel, PDF, HTML, and so on.

Usage of Oxygen Forensic Suite

The acquisition of an iOS device is simple and straightforward with Oxygen Forensic Suite 2014. The software helps you to connect a device in several mouse clicks and downloads all the available device information in just a few minutes.

To perform the acquisition of an iOS device using Oxygen Forensic Suite 2014, follow the steps provided:

1. Launch Oxygen Forensic Suite 2014 and click on the **Connect new device** button. You will be prompted to choose the connection mode, as shown in the following screenshot:

Oxygen Forensic Suite – the Connection Mode screen

2. Connect the iOS device to the computer using a USB cable and choose the **Auto device connection** mode. It detects the connected device and displays the device information, as shown in the following screenshot. You can also manually choose your device.

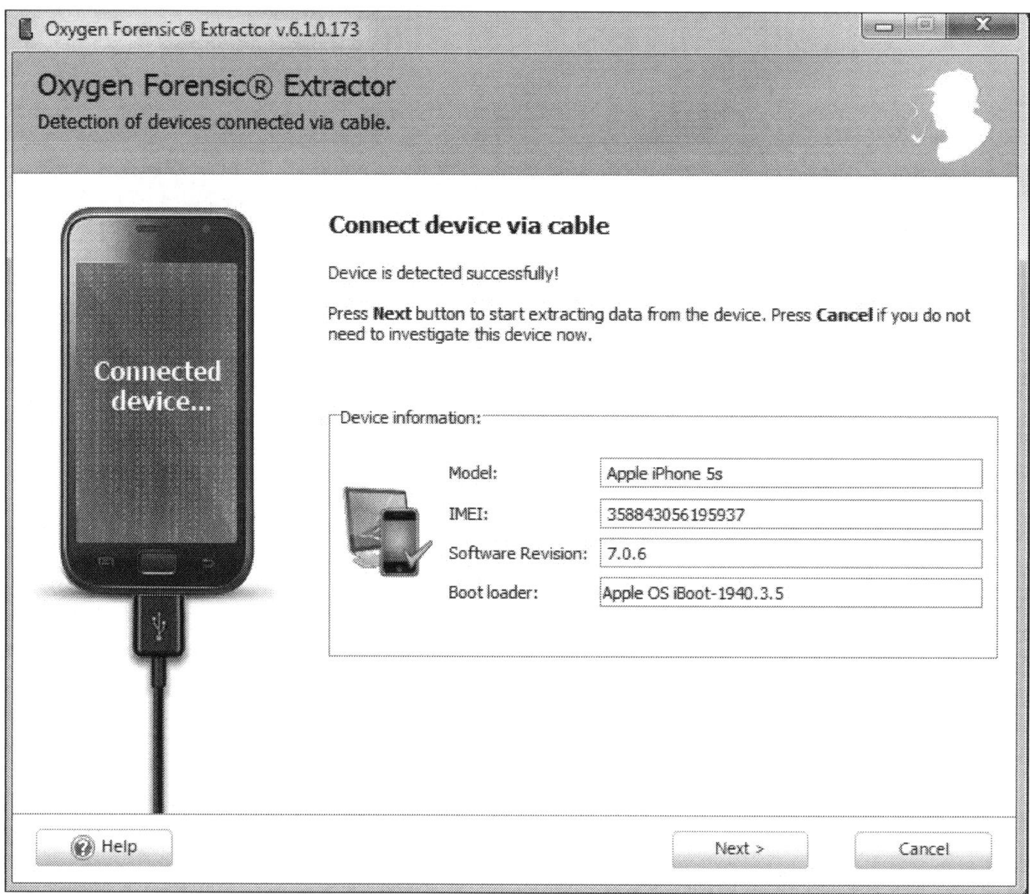

Oxygen Forensic Suite – the device information screen

3. Click on **Next**. It prompts you to fill in the information about the device and the case. Continuing further, it prompts you to select the data types to be extracted from the device, as shown in the following screenshot:

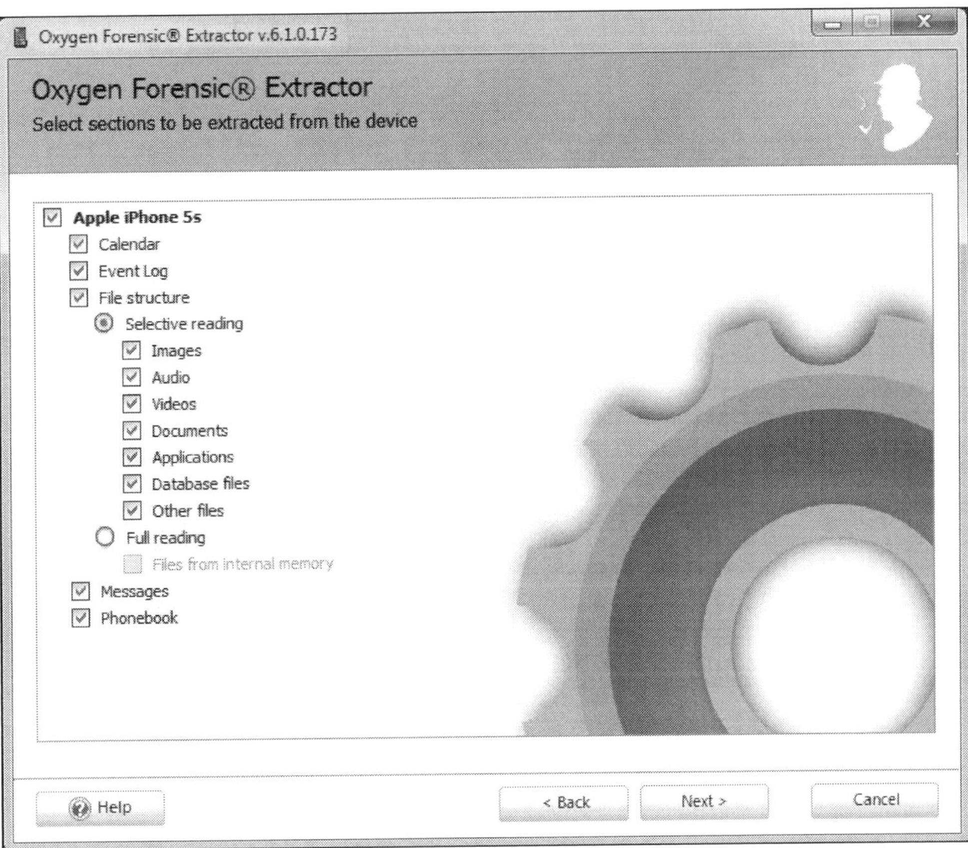

4. Click on **Next**. It extracts the data from the device and the process takes a few minutes depending on the amount of data stored on the device. Once the process is complete, the software displays a summary of the extracted data, as shown in the following screenshot:

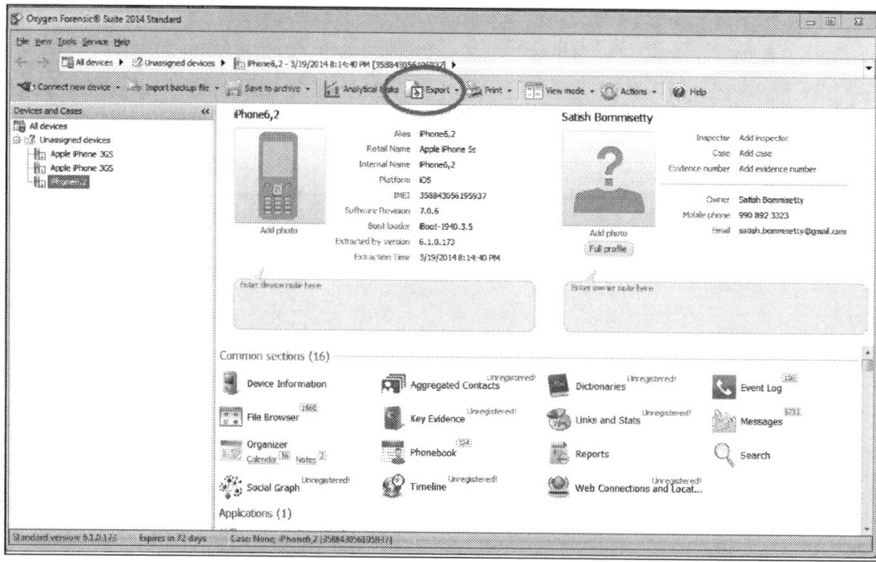

Oxygen Forensic Suite — the extracted data summary screen

5. After the download process is complete, you can use the automatic forensic report generation function and export the extracted data to a PDF file. The device data report appears as shown in the following screenshot. You can also open the device image in Oxygen for a manual look at the data.

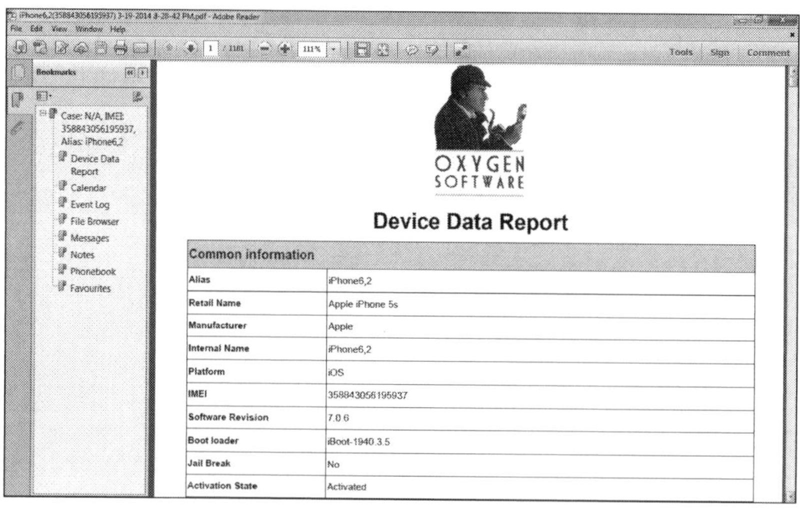

Device Data Report

Common information	
Alias	iPhone6,2
Retail Name	Apple iPhone 5s
Manufacturer	Apple
Internal Name	iPhone6,2
Platform	iOS
IMEI	358843056195937
Software Revision	7.0.6
Boot loader	iBoot-1940.3.5
Jail Break	No
Activation State	Activated

Oxygen Forensic Suite 2014 supported devices

Oxygen Forensic Suite 2014 Version 6.1 supports logical acquisition of all iOS devices. Keep in mind that access to newer devices may require the device to be unlocked or jailbroken.

Cellebrite UFED Physical Analyzer

As per the vendor, **Cellebrite UFED (Universal Forensic Extraction Device)** empowers law enforcement, anti-terrorism, and security organizations to capture critical forensic evidence from mobile phones, smartphones, PDAs, and portable handset varieties, including updates for newly released models. The tool enables forensically sound data extraction, decoding, and analysis techniques to obtain existing and deleted data from different mobile devices. As of February 2014, UFED supports data extraction from more than 5,320 mobile devices.

The Cellebrite UFED Physical Analyzer application can be used to perform physical and advanced logical acquisitions of iOS devices. Advanced logical acquisitions are the same as file system acquisitions in which access to the file system data is provided. Physical acquisition on iOS devices using the A5-A7 chip (iPhone 4s and newer) is not possible. Thus, the advanced logical acquisition method is the best support and will pull the most data from these devices if they are unlocked (even if they are not jailbroken). The application is available only for Windows platforms. Cellebrite also offers a 30-day free trial for the software. For more information, visit `http://www.cellebrite.com/mobile-forensics/products/applications/ufed-physical-analyzer`.

Features of Cellebrite UFED Physical Analyzer

The following are the features of Cellebrite UFED Physical Analyzer:

- Supports physical and advanced logical acquisition (file system acquisition)
- Extracts device keys required to decrypt raw disk images as well as keychain items
- Decrypts raw disk images and keychain items
- Reveals device passwords (not available for all locked devices)
- Allows to open an encrypted raw disk image file with a known password
- Supports passcode recovery attacks

- Advanced analysis and decoding of extracted applications data
- Reports generation in several popular formats—Microsoft Excel, PDF, HTML, and so on.
- Ability to dump the raw file system partition to import and examine it in another forensic tool

Usage of Cellebrite UFED Physical Analyzer

To perform the physical acquisition of an iPhone 4 and older devices with UFED Physical Analyzer, follow the steps provided. Note that physical acquisition is not supported for newer iOS devices (iPhone 4S and newer).

1. Launch UFED Physical Analyzer and navigate to the **Extract | iOS Device Extraction** menu. You will be prompted with the iOS device data extraction wizard, as shown in the following screenshot:

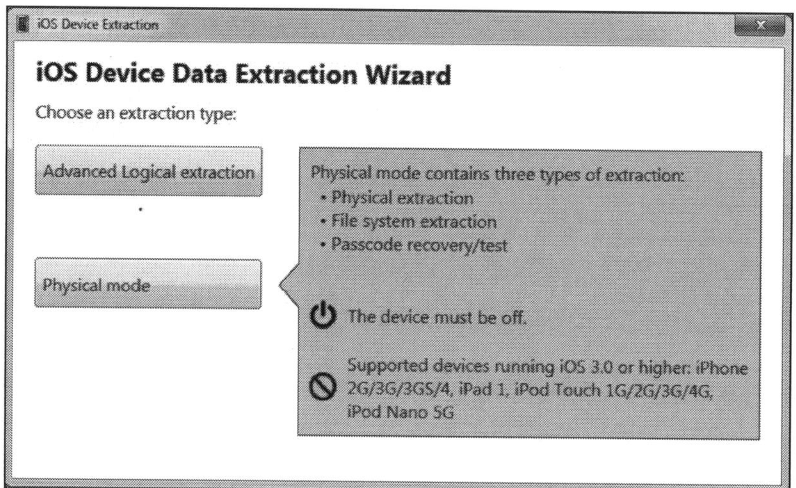

UFED Physical Analyzer—the iOS Device Data Extraction Wizard screen

2. Click on **Physical mode**. The first time you run iOS device extraction, you will be prompted to download and install the iOS support package.

3. Follow the instructions displayed on the screen to turn off the device and place it in the recovery mode. Once the tool detects the device in the recovery mode, it displays the device information, as shown in the following figure:

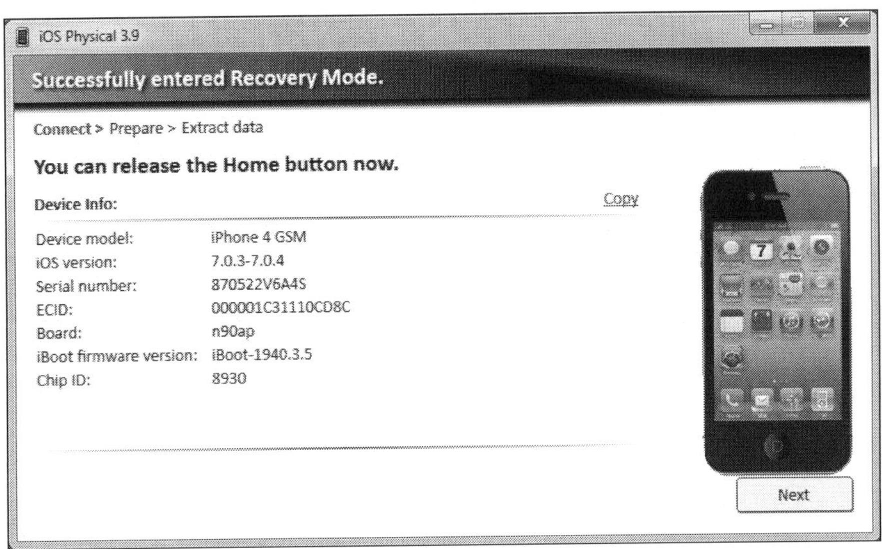

UFED Physical Analyzer—the device information screen

4. Click on **Next** and put the device in the DFU mode. When the device is detected in the DFU mode, the software loads the acquisition tools onto the device.
5. Once the device is ready for extraction, you will be prompted to choose the desired extraction type. Click on **Physical Extraction** and choose the partition you wish to extract and the location where you want to save the extraction.
6. Continue further and click on **Recover the passcode for me** to recover the passcode prior to the extraction.
7. Click on **Continue**. The tool extracts the file system image and decrypts it.

Supported devices

UFED Physical Analyzer Version 3.9 supported iOS devices are shown in the following table:

Model	iOS version	Physical acquisition	Logical acquisition
iPhone, iPhone 3G, iPod Touch 1, 2	iOS 1/2/3/4	Yes	Yes
iPhone 3GS iPod Touch 3 iPad 1	iOS 3/4/5	Yes	Yes
iPhone 4 iPod Touch 4	iOS 4/5/6/7	Yes	Yes
iPhone 4S, 5, 5C, 5S iPad 2, 3, 4, iPad mini, and iPod Touch 5	iOS 5/6/7	No	Yes

Paraben iRecovery Stick

As per the vendor, the **iRecovery Stick** contains specialized investigation software on a USB drive that allows anyone to investigate data on Apple iOS devices such as an iPhone, iPad, and iPod Touch. The iRecovery Stick acquires a user's data directly from the device or from iTunes backup files. The iRecovery Stick also recovers deleted data from SQLite databases and can recover data such as messages, contacts, call history, Internet history, and calendar events. Note that this is not a physical acquisition but is simply acquiring and parsing raw database files logically.

The iRecovery Stick costs $129 and works on Windows platforms. For better recovery, iRecovery Stick recommends turning off the antivirus software running on the computer. For more information, visit `http://www.paraben.com/irecovery-stick.html`.

Features of Paraben iRecovery Stick

The following are the features of Paraben iRecovery Stick:

- It supports logical acquisition
- It recovers deleted data from SQLite files
- It is easy to use and portable

- It is inconspicuous. It resembles a commonly used USB thumb drive, so it can be used as a spy device and no one would suspect that the device is used to recover data from an iPhone.
- It logs the recovery process based on the plugin activity and traffic across the communication port.
- It supports data analysis and reporting in several formats, such as Excel and PDF.

Usage of Paraben iRecovery Stick

The iRecovery Stick is a USB flash drive that contains the recovery software `iRecoveryStick.exe`.

To perform the acquisition of an iOS device using iRecovery Stick, follow these steps:

1. Connect the iOS device to the computer using a USB cable. Launch the iRecovery Stick software and click on the **Start Recovery** button. You will be prompted to choose the connected device, as shown in the following screenshot:

iRecovery Stick – the Choose connected device screen

2. Click on the device icon and it starts extracting the data from the device. The data extraction process takes a few minutes depending on the amount of data stored on the device.

3. Once the process is complete, the software displays a summary of extracted data, as shown in the following screenshot:

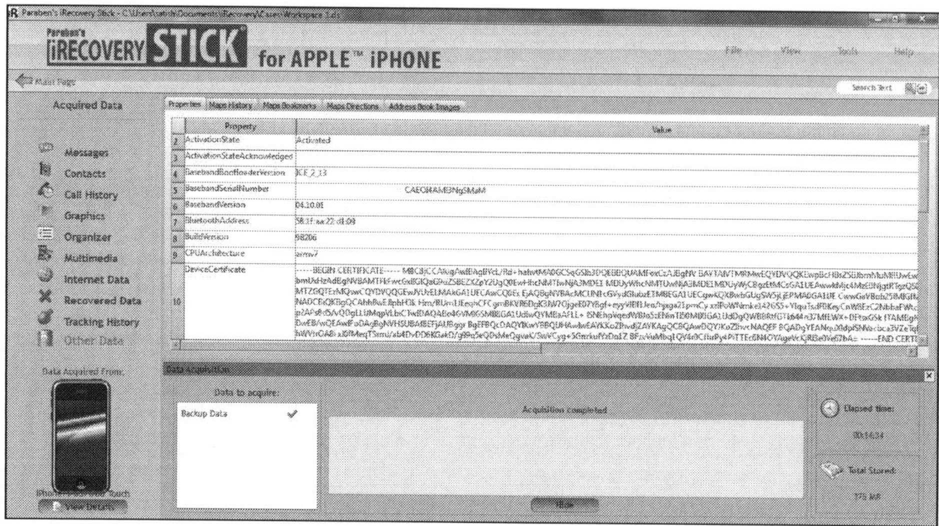

iRecovery Stick—the extracted data summary

Devices supported by Paraben iRecovery Stick

Paraben iRecovery Stick Version 3.5 supports logical acquisition of all iOS devices. The amount of data acquired will depend on how much data is present on the iOS device, whether the device was locked, and whether the device was jailbroken.

Open source or free methods

Several methods are available to acquire and analyze iOS devices for free. Most of these tools have been built by practitioners in mobile forensics who recognize the need for affordable solutions that work to obtain the same amount of data as commercial kits. Jon Zdziarski has developed several scripts, tools, and methods to acquire data from iOS devices. Some of his methods such as physical acquisition scripts are restricted to law enforcement. Zdziarksi released his instructions to acquire data from iOS devices and this can be read at `http://www.zdziarski.com/blog/wp-content/uploads/2013/05/iOS-Forensic-Investigative-Methods.pdf`.

There are other tools that exist so you can logically acquire and analyze iOS device images and backup files. Some of these tools include iFunBox, iExplorer, iBackupBot, and more. Make sure that you test these tools before relying on them for a forensic investigation. Again, they are either free or request a donation for use. They are developed by the community for examiners to use. They often do not go through rigorous amounts of testing and validation and may miss data that can be manually extracted by the examiner. It is the examiner's responsibility to learn the tool, test it, and know its flaws in order to recover all of the available data.

Summary

Forensic tools are helpful for an investigator as they not only save time but also make the process a lot easier. This chapter introduced you to several available iOS forensic tools and included the steps to perform acquisition of an iOS device. Examiners should take further steps to validate and understand each tool that might be used as part of an investigation. In the next chapter, we will dive into Android forensics and provide information on what Android is, how the devices store data, and how to access the files and applications that are required for forensic examinations.

7
Understanding Android

Before we take a dive into the ocean of Android let us first spend some time discussing the evolution of Android or what we call **The Android Story**. Back in 2005, Google started investing money in start-up companies, which it thought would be profitable in the future. Android Inc., founded in 2003 by Andy Rubin, Rich Miner, Nick Sears, and Chris White, was one such company acquired by Google that later turned out to be the best deal ever. During its first two years, Android Inc. operated under secrecy. It described itself as a company "making software for mobile phones". Rubin later stayed with Google to pioneer Android as an operating system that revolutionized the way mobile handsets operate. With this acquisition it was clear that Google was eyeing the mobile phone market. At Google, Rubin, along with his team, developed a powerful and flexible operating system built on a Linux kernel. There were speculations all over about what Google was trying to do. Some reported that Google was trying to incorporate search and other applications into mobile handsets. A few others reported that Google was developing its own mobile handset. Finally in 2007, **Open Handset Alliance (OHA)**, a group of technology companies, device manufacturers, chipset makers, and wireless carriers, was formed with the main objective of proposing open standards for the mobile platform. Together they developed **Android**, the first open and free mobile platform built on Linux kernel 2.6. Later in 2008, HTC Dream was released which was the first phone to run the Android operating system. After that, it was a dream run for Android, with its market share increasing exponentially over the next few years. A breakdown on the history of Android can be found at `http://www.xcubelabs.com/the-android-story.php`. Several versions of its Linux-based operating system have been released in alphabetical order.

The version history of Android can be found at `http://faqoid.com/advisor/android-versions.php`, an overview of which is shown in the following table:

Version	Version name	Release year
Android 1.0	Apple pie	2008
Android 1.1	Banana bread	2009
Android 1.5	Cupcake	2009
Android 1.6	Donut	2009
Android 2.0	Eclair	2009
Android 2.2	Froyo	2010
Android 2.3	Gingerbread	2010
Android 3.0	Honeycomb	2011
Android 4.0	Ice Cream Sandwich	2011
Android 4.1	Jelly Bean	2012
Android 4.4	KitKat	2013

The Android model

To effectively understand the forensic concepts of Android, it would be helpful to have a basic understanding of the Android architecture. Just like a computer, any computing system that interacts with the user and performs complicated tasks requires an operating system to handle the tasks effectively. This operating system (whether it's a desktop operating system or a mobile phone operating system) takes the responsibility to manage the resources of the system and to provide a way for the applications to talk to the hardware or physical components to accomplish certain tasks. Android is currently the most popular mobile operating system designed to power mobile devices. You can find out more about this at `http://developer.android.com/about/index.html`. Android is open source and the code is released under Apache license. Practically, this means anyone (especially device manufacturers) can access it, freely modify it, and use the software according to the requirements of any device. This is one of the primary reasons for its wide acceptance. Notable players that use Android include Samsung, HTC, Sony, LG, and so on.

As with any other platform, Android consists of a stack of layers running one above the other. To understand the Android ecosystem, it's essential to have a basic understanding of what these layers are and what they do. The following figure summarizes the various layers involved in the Android software stack (`https://viaforensics.com/wp-content/uploads/2009/08/Android-Forensics-Andrew-Hoog-viaForensics.pdf`):

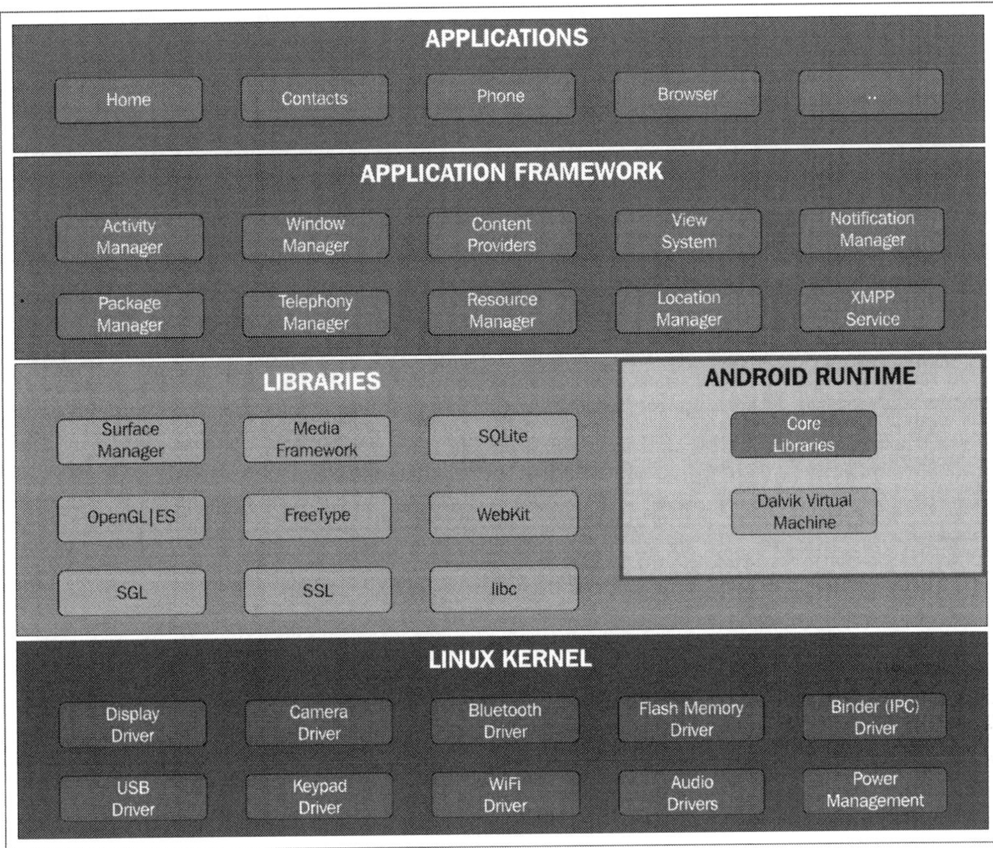

Android architecture

Each of these layers performs several operations that support specific operating system functions (`http://www.android-app-market.com/android-architecture.html`). Each layer provides services to the layers lying on top of it.

The Linux kernel layer

Android OS is built on top of the Linux kernel with some architectural changes made by Google. There are several reasons for choosing the Linux kernel. Most importantly, Linux is a portable platform that can be compiled easily on different hardware. The kernel acts as an abstraction layer between the software and hardware present on the device. Consider the case of a camera click. What happens when you click a photo using the camera button on your device? At some point, the hardware instruction (pressing a button) has to be converted to a software instruction (to take a picture and store it in the gallery). The kernel contains drivers to facilitate this process. When the user clicks on the button, the instruction goes to the corresponding camera driver in the kernel, which sends the necessary commands to the camera hardware, similar to what occurs when a key is pressed on a keyboard. In simple words, the drivers in the kernel command control the underlying hardware. As shown in the preceding figure, the kernel contains drivers related to Wi-Fi, Bluetooth, USB, audio, display, and so on.

The Linux kernel is responsible for managing the core functionality of Android, such as process management, memory management, security, and networking. Linux is a proven platform when it comes to security and process management. Android has taken leverage of the existing Linux open source OS to build a solid foundation for its ecosystem. Each version of Android has a different version of the underlying Linux kernel. The current KitKat Android version is rumored to use Linux kernel 3.8 (`http://www.phonearena.com/news/Android-4.4-KitKat-update-release-date-features-and-rumors_id47661`).

Libraries

The next layer in the Android architecture consists of Android's native libraries. The libraries are written in the C or C++ language and help the device to handle different kinds of data. For example, the SQLite libraries are useful for storing and retrieving the data from a database. Other libraries include Media Framework, WebKit, Surface Manager, SSL, and so on. The Media Framework library acts as the main interface to provide a service to the other underlying libraries. The WebKit library provides web pages in web browsers and the surface manager maintains the graphics. In the same layer, we have Android Runtime, which consists of Dalvik virtual machine (DVM) and core libraries. The Android runtime is responsible for running applications on Android devices. The term "runtime" refers to the lapse in time from when an application is launched until it is shut down.

Dalvik virtual machine

All the applications that you install on the Android device are written in the Java programming language. When a Java program is compiled, we get bytecode. JVM is a virtual machine (a virtual machine is an application that acts as an operating system, that is, it is possible to run a Windows OS on a Mac or vice versa by using a virtual machine) that can execute this bytecode. But Android uses something called **Dalvik virtual machine (DVM)** to run its applications.

DVM runs Dalvik bytecode, which is Java bytecode converted by the Dex compiler (`http://markfaction.wordpress.com/2012/07/15/stack-based-vs-register-based-virtual-machine-architecture-and-the-dalvik-vm/`). Thus, the `.class` files are converted to dex files using the dx tool. Dalvik bytecode when compared with Jave bytecode is more suitable for low-memory and low-processing environments. Also, note that JVM's bytecode consists of one or more `.class` files depending on the number of Java files that are present in an application, but Dalvik bytecode is composed of only one dex file. Each Android application runs its own instance of Dalvik virtual machine. This is a crucial aspect of Android security and will be addressed in detail in *Chapter 8, Android Forensic Setup and Pre Data Extraction Techniques*. The following figure provides an insight into how Android's DVM differs from Java's JVM.

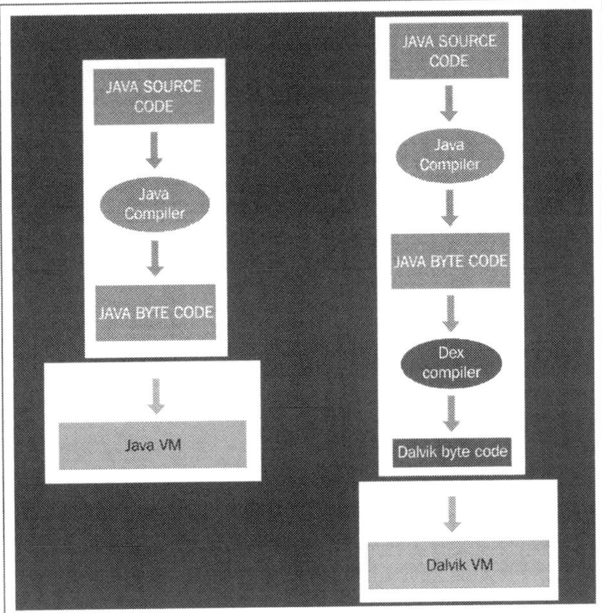

JVM versus DVM

The application framework layer

The application framework is the layer responsible for handling the basic functioning of a phone, such as resource management, handling calls, and so on. This is the block with which the applications installed on the device directly talk to it. The following are some of the important blocks in the application framework layer:

- **Telephony Manager**: This block manages all the voice calls
- **Content Provider**: This block manages the sharing of data between different applications
- **Resource Manager**: This block helps manage various resources used in applications

The applications layer

This is the topmost layer where the user can interact directly with the device. There are two kinds of applications—preinstalled applications and user-installed applications. Preinstalled applications, such as Dialer, Web Browser, Contacts, and more come along with the device. User-installed applications can be downloaded from different places, such as Google Play Store, Amazon Marketplace, and so on. Everything that you see on your phone (contacts, mail, camera, and so on) is an application.

Android security

Android was designed with a specific focus on security. Android as a platform offers and enforces certain features that safeguard the user data present on the mobile through multilayered security. There are certain safe-defaults that will protect the user and certain offerings that can be leveraged by the development community to build secure applications. The following are issues which are kept in mind while incorporating the Android security controls:

- Protecting user-related data
- Safeguarding the system resources
- Making sure one application cannot access the data of another application

The next few concepts help us understand more about Android's security features and offerings. A detailed explanation on Android security can be found at `http://source.android.com/devices/tech/security/`.

Secure kernel

Linux has evolved as a trusted platform over the years, and Android has leveraged this fact by using it as its kernel. The user-based permission model of Linux has in fact worked well for Android. As mentioned earlier, there is a lot of specific code built into the Linux kernel. With each Android version release, the kernel version has also changed. The following table shows Android versions and their corresponding kernel versions:

Android version	Linux kernel version
1	2.6.25
1.5	2.6.27
1.6	2.6.29
2.2	2.6.32
2.3	2.6.35
3	2.6.36
4	3.0.1
4.1	3.0.31
4.2	3.4.0
4.2	3.4.39
4.4	3.8

Linux kernel versions used in Android

The permission model

As shown in the following screenshot, any Android application must be granted permissions to access sensitive functionality, such as the Internet, dialer, and so on, by the user. This provides an opportunity for the user to know in advance what functionality on the device the application is trying to access. Simply put, it requires the user's permission to perform any kind of malicious activity (stealing data, compromising the system, and so on).

This model helps the user to prevent attacks, but if the user is unaware and gives away a lot of permissions, it leaves them in trouble (remember when it comes to installing malware on any device, the weakest link is always the user).

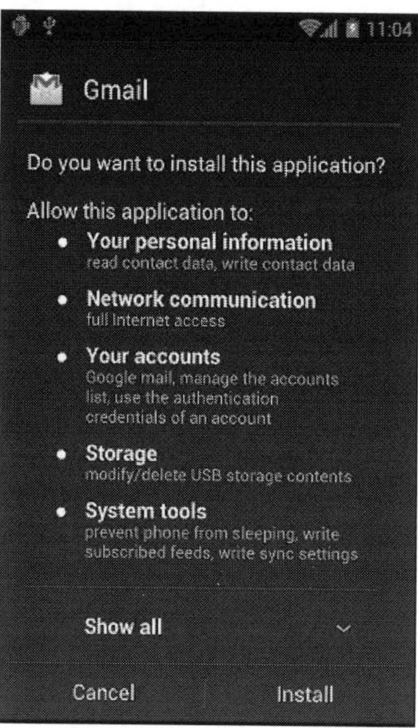

The permission model in Android

Application sandbox

In Linux systems, each user is assigned a unique user ID (UID), and users are segregated so that one user can access the data of another user. However, all applications under a particular user are run with the same privileges. Similarly in Android, each application runs as a unique user. In other words, a UID is assigned to each application and is run as a separate process. This concept ensures an application sandbox at the kernel level. The kernel manages the security restrictions between the applications by making use of existing Linux concepts, such as UID and GID. If an application attempts to do something malicious, say to read the data of another application, this is not permitted as the application does not have the user privileges. Hence, the operating system protects an application from accessing the data of another application.

Secure interprocess communication

Android offers a secure interprocess communication through which one's activity in an application can send messages to another activity in the same application or a different application. To achieve this, Android provides interprocess communication (IPC) mechanisms: intents, services, content providers, and so on.

Application signing

It is mandatory that all the installed applications be digitally signed. Developers can place their applications in Google's Play Store only after signing the applications. The private key with which the application is signed is held by the developer. Using the same key, a developer can provide updates to their application, share data between the applications, and so on.

Android file hierarchy

In order to perform forensic analysis on any system (desktop or mobile), it's important to understand the underlying file hierarchy. A basic understanding of how Android organizes its data in files and folders helps a forensic analyst narrow down their research to specific issues. Just like any other operating system, Android uses several partitions. This chapter provides an insight into some of the most significant partitions and the content stored in them.

It's worth mentioning again that Android uses the Linux kernel. Hence, if you are familiar with Unix-like systems, you will very well understand the file hierarchy in Android. For those who are not very well acquainted with the Linux model, here is some basic information: in Linux, the file hierarchy is a single tree with the top of the tree being denoted as / (called the "root"). This is different from the concept of organizing files in drives (as with Windows). Whether the file system is local or remote, it will be present under the root. The Android file hierarchy is a customized version of this existing Linux hierarchy. Based on the device manufacturer and the underlying Linux version, the structure of this hierarchy may have a few insignificant changes. The following is a list of important folders that are common to most Android devices. Some of the folders listed are only visible through root access.

- `/boot`: As the name suggests, this partition has the information and files required for the phone to boot. It contains the kernel and RAM disk, and so without this partition the phone cannot start its processes. Data residing in RAM is rich in value and should be captured during a forensic acquisition.

- `/system`: This partition contains system-related files other than kernel and RAM disk. This folder should never be deleted as that will make the device unbootable. The contents of this partition can be viewed by using the following command:

```
shell@Android:/ $ cd /system
cd /system
shell@Android:/system $ ls
ls
CSCVersion.txt
SW_Configuration.xml
app
bin
build.prop
cameradata
csc
csc_contents
etc
fonts
framework
hdic
lib
media
recovery-from-boot.p
sipdb
tts
usr
vendor
voicebargeindata
vsc
wakeupdata
wallpaper
xbi
```

- `/recovery`: This is designed for backup purposes and allows the device to boot into the recovery mode. In the recovery mode, you can find tools to repair your phone installation.

- `/data`: This is the partition that contains the data of each application. Most of the data belonging to the user, such as the contacts, SMS, and dialed numbers, is stored in this folder. This folder has significant importance from a forensic point of view as it holds valuable data. The contents of the `data` folder can be viewed using the following command:

```
C:\Android-sdk-windows\platform-tools>adb.exe shell
root@Android:/ # cd /data
cd /data
root@Android:/data # ls
ls
anr
app
app-private
backup
camera
dalvik-cache
data
dontpanic
drm
local
lost+found
misc
property
resource-cache
system
system.notfirstrun
user
```

- `/cache`: This is the folder used to store frequently accessed data and some of the logs for faster retrieval. The `cache` partition is also important to the forensic investigation as the data residing here may no longer be present in the `/data` partition.
- `/misc`: As the name suggests, this folder contains information about miscellaneous settings. These settings mostly define the state of the device, that is On/Off. Information about hardware settings, USB settings, and so on, can be accessed from this folder.

Android file system

Understanding the file system is one essential part of forensic methodologies. Knowledge about properties and the structure of a file system proves to be useful during forensic analysis. File system refers to the way data is stored, organized, and retrieved from a volume. A basic installation may be based on one volume split into several partitions; here each partition can be managed by a different file system. As is true in Linux, Android utilizes mount points and not drives (that is `C:` or `E:`). Each file system defines its own rules for managing the files on the volume. Depending on these rules, each file system offers a different speed for file retrieval, security, size, and so on. Linux uses several file systems, and so does Android. From a forensic point of view, it's important to understand what file systems are used by Android and to identify the file systems that are of significance to the investigation. For example, the file system that stores the user's data is of primary concern to us as against a file system used to boot the device.

Viewing file systems on an Android device

The file systems supported by the Android kernel can be determined by checking the contents of the file `filesystems` in the `proc` folder. The content of this file can be viewed by using the following command:

```
shell@Android:/ $ cat /proc/filesystems
cat /proc/filesystems
nodev   sysfs
nodev   rootfs
nodev   bdev
nodev   proc
nodev   cgroup
nodev   tmpfs
nodev   binfmt_misc
nodev   debugfs
nodev   sockfs
nodev   usbfs
nodev   pipefs
nodev   anon_inodefs
nodev   devpts
```

```
        ext2
        ext3
        ext4
nodev   ramfs
        vfat
        msdos
nodev   ecryptfs
nodev   fuse
        fuseblk
nodev   fusectl
        exfat
```

In the preceding output, the first column tells us whether the file system is mounted on the device. The ones with the `nodev` property are not mounted on the device. The second column lists all the file systems present on the device. A simple mount command displays different partitions available on the device, as follows:

```
shell@Android:/ $ mount
mount
rootfs / rootfs ro,relatime 0 0
tmpfs /dev tmpfs rw,nosuid,relatime,mode=755 0 0
devpts /dev/pts devpts rw,relatime,mode=600 0 0
proc /proc proc rw,relatime 0 0
sysfs /sys sysfs rw,relatime 0 0
none /acct cgroup rw,relatime,cpuacct 0 0
tmpfs /mnt/asec tmpfs rw,relatime,mode=755,gid=1000 0 0
tmpfs /mnt/obb tmpfs rw,relatime,mode=755,gid=1000 0 0
none /dev/cpuctl cgroup rw,relatime,cpu 0 0
/dev/block/mmcblk0p9 /system ext4 ro,noatime,barrier=1,data=ordered 0
0
/dev/block/mmcblk0p3 /efs ext4
rw,nosuid,nodev,noatime,barrier=1,journal_async_commit,data=ordered 0
0
/dev/block/mmcblk0p8 /cache ext4
rw,nosuid,nodev,noatime,barrier=1,journal_async
_commit,data=ordered 0 0
/dev/block/mmcblk0p12 /data ext4
rw,nosuid,nodev,noatime,barrier=1,journal_async
```

```
_commit,data=ordered,noauto_da_alloc,discard 0 0
/sys/kernel/debug /sys/kernel/debug debugfs rw,relatime 0 0
/dev/fuse /storage/sdcard0 fuse
rw,nosuid,nodev,noexec,relatime,user_id=1023,gro
up_id=1023,default_permissions,allow_other 0 0
```

The next few sections provide a brief overview of the important file systems.

The root file system (`rootfs`) is one of the main components of Android and contains all the information required to boot the device. When the device starts the boot process, it needs access to many core files and thus mounts the root file system. As shown in the preceding mount command-line output, this file system is mounted at / (root folder). Hence, this is the file system on which all the other file systems are slowly mounted. If this file system is corrupt, the device cannot be booted.

The `sysfs` file system mounts the `/sys` folder, which contains information about the configuration of the device. The following output shows various folders under the `sys` directory in an Android device:

```
shell@Android:/ $ cd /sys
cd /sys
shell@Android:/sys $ ls
ls
block
bus
class
dev
devices
firmware
fs
kernel
module
power
```

Since the data present in these folders is mostly related to configuration, this is not usually of much significance to a forensic investigator. But there could be some circumstances where we might want to check if a particular setting was enabled on the phone, and analyzing this folder could be useful under such conditions. Note that each folder consists of a large number of files. Capturing this data through forensic acquisition is the best method to ensure this data is not changed during examination.

The devpts file system presents an interface to the terminal session on an Android device. It is mounted at /dev/pts. Whenever a terminal connection is established, for instance, when an adb shell is connected to an Android device, a new node is created under /dev/pts. The following is the output showing this when the adb shell is connected to the device:

```
shell@Android:/ $ ls -l /dev/pts
ls -l /dev/pts
crw------- shell    shell    136,   0 2013-10-26 16:56 0
```

The cgroup file system stands for control groups. Android devices use this file system to track their job. They are responsible for aggregating the tasks and keeping track of them. This data is generally not very useful during forensic analysis.

The proc file system contains information about kernel data structures, processes, and other system-related information under the /proc directory. For instance, the /sys directory contains files related to kernel parameters. Similarly, /proc/filesystems displays the list of available file systems on the device. The following command shows all information about the CPU of the device:

```
shell@Android:/ $ cat /proc/cpuinfo
cat /proc/cpuinfo
Processor       : ARMv7 Processor rev 0 (v7l)
processor       : 0
BogoMIPS        : 1592.52

processor       : 3
BogoMIPS        : 2786.91

Features        : swp half thumb fastmult vfp edsp neon vfpv3 tls
CPU implementer : 0x41
CPU architecture: 7
CPU variant     : 0x3
CPU part        : 0xc09
CPU revision    : 0

Chip revision   : 0011
Hardware        : SMDK4x12
Revision        : 000c
Serial          : ****************
```

Similarly, there are many other useful files that provide valuable information when you traverse through them.

The `tmpfs` file system is a temporary storage facility on the device that stores the files in RAM (volatile memory). The main advantage of using RAM is faster access and retrieval. But once the device is restarted or switched off, this data will not be accessible anymore. Hence, it's important for a forensic investigator to examine the data in RAM before a device reboot happens or extract the data via RAM acquisition methods.

Extended File System – EXT

Extended File System (EXT), which was introduced in 1992 specifically for the Linux kernel, was one of the first file systems and used the virtual file system. EXT2, EXT3, and EXT4 are the subsequent versions. Journaling is the main advantage of EXT3 over EXT2. With EXT3, in case of an unexpected shutdown, there is no need to verify the file system. the EXT4 file system, the fourth extended file system, has gained significance with mobile devices implementing dual-core processors. The YAFFS2 file system is known to have a bottleneck on dual-core systems. With the Gingerbread version of Android, the YAFFS file system was swapped for EXT4. The following are the mount points that use EXT4 on Samsung Galaxy S3 mobile:

```
/dev/block/mmcblk0p9 /system ext4 ro,noatime,barrier=1,data=ordered 0
0

/dev/block/mmcblk0p3 /efs ext4
rw,nosuid,nodev,noatime,barrier=1,journal_async_commit,data=ordered 0
0

/dev/block/mmcblk0p8 /cache ext4
rw,nosuid,nodev,noatime,barrier=1,journal_async_commit,data=ordered 0
0

/dev/block/mmcblk0p12 /data ext4
rw,nosuid,nodev,noatime,barrier=1,journal_async_commit,data=ordered,n
oauto_da_alloc,discard 0 0
```

VFAT is an extension to the FAT16 and FAT32 file systems. Microsoft's FAT32 file system is supported by most Android devices. It is supported by almost all the major operating systems, including Windows, Linux, and Mac OS. This enables these systems to easily read, modify, and delete the files present on the FAT32 portion of the Android device. Most of the external SD cards are formatted using the FAT32 file system. Observe the following output, which shows that the mount points `/sdcard` and `/secure/asec` use the VFAT file system.

```
shell@Android:/sdcard $ mount
mount
rootfs / rootfs rw 0 0
tmpfs /dev tmpfs rw,nosuid,relatime,mode=755 0 0
devpts /dev/pts devpts rw,relatime,mode=600,ptmxmode=000 0 0
proc /proc proc rw,relatime 0 0
sysfs /sys sysfs rw,relatime 0 0
tmpfs /mnt/asec tmpfs rw,relatime,mode=755,gid=1000 0 0
tmpfs /mnt/obb tmpfs rw,relatime,mode=755,gid=1000 0 0
/dev/block/nandd /system ext4
rw,nodev,noatime,user_xattr,barrier=0,data=ordered 0 0
/dev/block/nande /data ext4
rw,nosuid,nodev,noatime,user_xattr,barrier=0,journal_checksum,data=or
dered,noauto_da_alloc 0 0
/dev/block/nandh /cache ext4
rw,nosuid,nodev,noatime,user_xattr,barrier=0,journal_checksum,data=or
dered,noauto_da_alloc 0 0
/dev/block/vold/93:64 /mnt/sdcard vfat
rw,dirsync,nosuid,nodev,noexec,relatime,uid=1000,gid=1015,fmask=0702,
dmask=0702,allow_utime=0020,codepage=cp437,iocharset=ascii,shortname=
mixed,utf8,errors=remount-ro 0 0
/dev/block/vold/93:64 /mnt/secure/asec vfat
rw,dirsync,nosuid,nodev,noexec,relatime,uid=1000,gid=1015,fmask=0702,
dmask=0702,allow_utime=0020,codepage=cp437,iocharset=ascii,shortname=
mixed,utf8,errors=remount-ro 0 0
tmpfs /mnt/sdcard/.Android_secure tmpfs ro,relatime,size=0k,mode=000
0 0
/dev/block/dm-0 /mnt/asec/com.kiloo.subwaysurf-1 vfat
ro,dirsync,nosuid,nodev,relatime,uid=1000,fmask=0222,dmask=0222,codep
age=cp437,iocharset=ascii,
shortname=mixed,utf8,errors=remount-ro 0 0
```

Yet Another Flash File System 2 (YAFFS2) is an open source, single-threaded file system released in 2002. It is mainly designed to be fast when dealing with NAND flash. YAFFS2 utilizes OOB (out of band) and that is often not captured or decoded correctly during forensic acquisition, which makes analysis difficult. This will be discussed more in *Chapter 9, Android Data Extraction Techniques*. YAFFS2 was the most popular release at one point and is still widely used in Android devices. YAFFS2 is a log-structured file system. Data integrity is guaranteed even in case of sudden power outage. In 2010, there was an announcement stating that in releases after Gingerbread, devices were going to move from YAFFS2 to EXT4. Currently YAFFS2 is not supported in newer kernel versions, but certain mobile manufacturers might still continue to support it.

Flash Friendly File System (F2FS) was released in February 2013 to support Samsung devices running the Linux 3.8 kernel (`http://www.linux.org/threads/flash-friendly-file-system-f2fs.4477/`). F2FS relies on log-structured methods that optimize NAND flash memory. The offline support features are a highlight of this file system. Yet, the file system is still transient and being updated.

Robust File System (RFS) supports NAND flash memory on Samsung devices. RFS can be summarized as a FAT16 (or FAT32) file system where journaling is enabled through a transaction log. Many users complain that Samsung should stick with EXT4. RFS has been known to have lag times that slow down the features of Android.

Summary

Understanding the underlying features, file systems, and capabilities of an Android device proves useful in a forensic investigation. Unlike iOS, several variants of Android exist as many devices run the Android operating system and each may have different file systems and unique features. The fact that Android is open and customizable also changes the playing field of digital forensics. A forensic examiner must be prepared to expect the unexpected when handling an Android device. In the next chapter, we will discuss methods for accessing the data stored on Android devices.

8

Android Forensic Setup and Pre Data Extraction Techniques

Having an established forensic environment before the start of an examination is important as it ensures that the data is protected while the examiner maintains control of the workstation. This chapter will explain the process and considerations when setting up a digital forensic examination environment. It is paramount that the examiner maintains control of the forensic environment at all times. This prevents the introduction of cross contaminants that could effect the forensic investigation. This chapter aims to cover the minimum basic requirements that should be in place to start a forensic investigation of an Android mobile device.

A forensic environment setup

Setting up a proper lab environment is an essential part of a forensic process. Android forensic setup usually involves the following:

- Start with a fresh or forensically sterile computer environment. This means that other data is not present on the system or is contained in a manner that it cannot contaminate the present investigation.
- Install basic software necessary to connect to the device. Android forensic tools and methodologies will work on Windows, Linux, and OS X platforms.

- Obtain access to the device. An examiner must be able to enable settings or bypass them in order to allow the data to be extracted from the Android device.
- Issue commands to the device through the methods defined in this chapter and in *Chapter 9, Android Data Extraction Techniques*.

The following sections provide guidance on setting up a basic Android forensic workstation.

Android Software Development Kit

The Android **Software Development Kit** (**SDK**) helps the development world to build, test, and debug applications to run on Android. This is achieved by providing necessary tools to create the applications. But along with this, it also provides valuable documentation and other tools that can be of great help during the investigation of an Android device. A good understanding of the Android SDK will help you to get to grips with the particulars of a device and the data on the device.

The Android SDK consists of software libraries, APIs, tools, emulators, and other reference material. It can be downloaded for free from `http://developer.android.com/`. During a forensic investigation, the SDK helps connect to and access the data on the Android device. The Android SDK is updated very frequently so it's important to verify that your workstation also remains updated. The Android SDK can run on Windows, Linux, and OS X.

Android SDK installation

A working installation of The Android SDK is a must during the investigation of a forensic device. Most websites recognize the operating system on the computer and will prompt you to download the correct Android SDK. The following is a step-by-step procedure to install the Android SDK on a Windows 7 machine:

1. Before you install the Android SDK, make sure your system has Java Development Kit installed because the Android SDK relies on **Java SE Development Kit** (**JDK**). JDK can be downloaded from `http://www.oracle.com/technetwork/java/javase/downloads/index.html`.

2. Download the latest version of the Android SDK from `http://developer.android.com/`. The installer version of the SDK is recommended for this purpose.
3. Run the installer file, which was downloaded in step 2. A wizard window will be shown, as seen in the following screenshot. After this, run through the routine **Next** steps that you encounter.

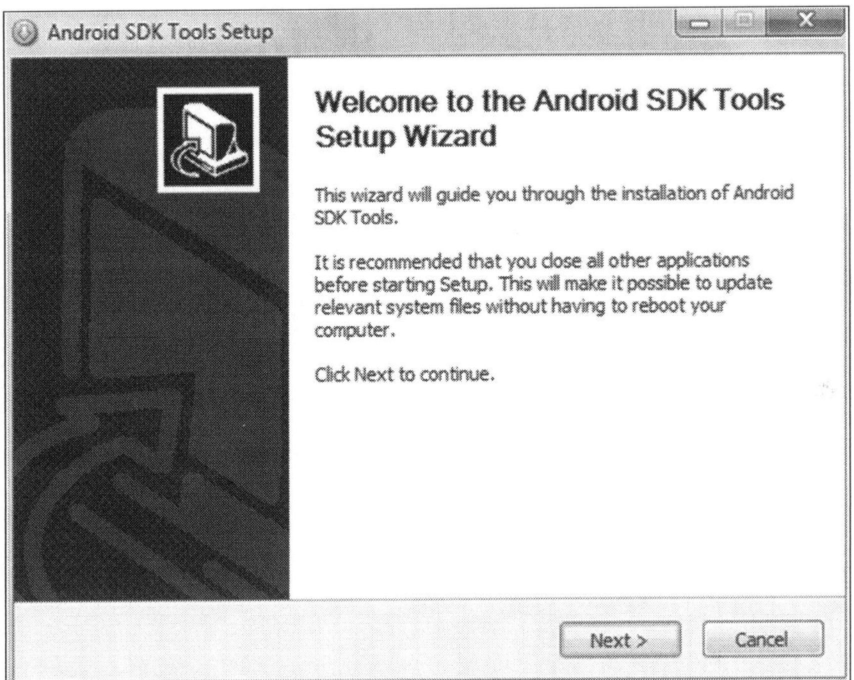

Android SDK Tools setup wizard

4. The installation location is the user's choice and must be remembered for future access. In this example, we will install it in the `C:\` folder. Click on the **Install** button and choose the location (say, `C:\android-sdk`). The necessary files will be extracted to this folder.

5. Open the directory (`C:\android-sdk`) and double-click on `SDK Manager.exe` to begin the update process. Make sure that you select Android SDK Platform tools and any one release platform version of Android as shown in the following screenshot. Some of the items in the list are chosen by default. For instance, it is necessary to install the USB driver in order to work with Android devices in Windows. In our example, **Google USB Driver** is selected. Similarly, you can find other items under the **Extras** section. Accept the license and install it, as shown in the following screenshot:

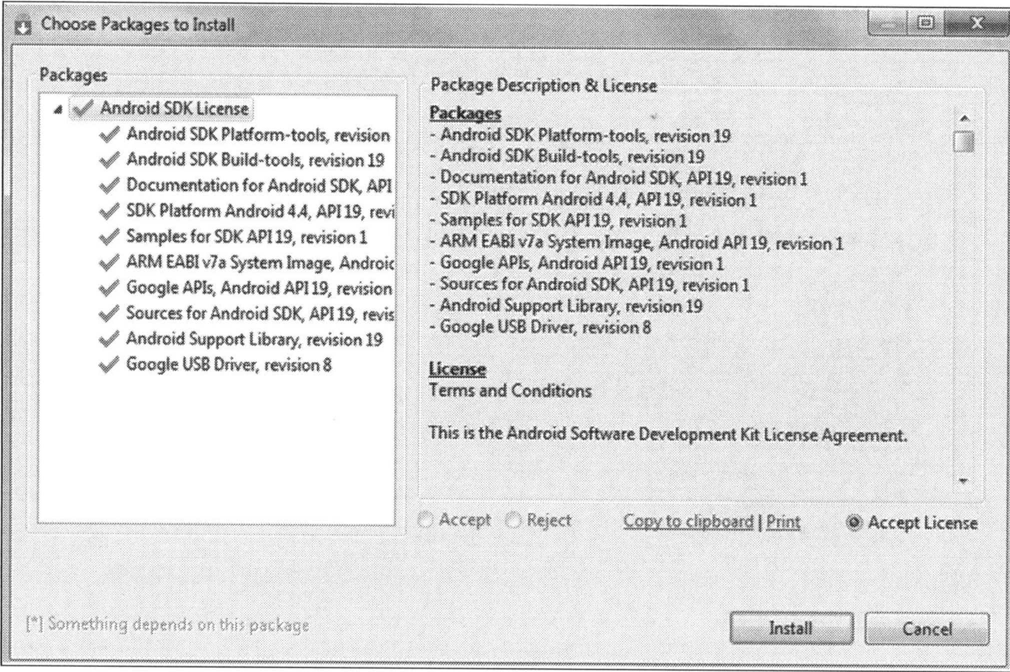

Android SDK License

This completes the Android SDK installation and you can update the system's environment variables (Path) by pointing to the executable files. The installation of the Android SDK on OS X and Linux may vary. Make sure that you follow all the steps provided with the SDK download for full functionality.

Android Virtual Device

Once the Android SDK is installed along with the release platform, you can create an **Android Virtual Device** (also called an emulator/**AVD**), which is often used by developers when creating new applications. However, an emulator has significance from a forensic perspective too. Emulators are useful when trying to understand how applications behave and execute on a device. This could be helpful to confirm certain findings that are unearthed during a forensic investigation. Also, while working on a device which is running on an older platform, you can design an emulator with the same platform. Furthermore, before installing a forensic tool on a real device, the emulator can be used to find out how a forensic tool works and changes content on an Android device. To create a new AVD (on the Windows workstation), perform the following steps:

1. Open the command prompt (`cmd.exe`). To start the AVD manager from the command line, navigate to the path where the SDK is installed and call the `android` tool with the `avd` option as shown in the following command line. This would automatically open the AVD manager.

   ```
   C:\android-sdk\tools>android avd
   ```

Alternately, the AVD manager can also be started using the graphical AVD manager. To start this, navigate to the location where the SDK is installed (`C:\android-sdk`) in our example and double-click on **AVD Manager**.

The **Android Virtual Device Manager** window is as shown in the following screenshot:

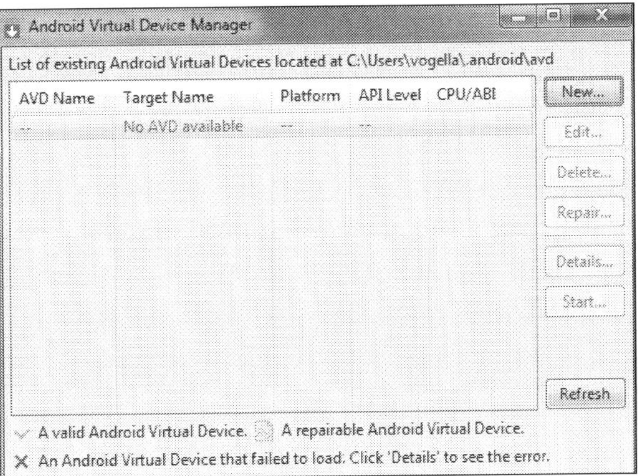

Android Virtual Device Manager

2. Click on **New** in the AVD Manager window to create a new virtual device. Click on **Edit** to change the configuration of an existing virtual device as shown in the following screenshot:

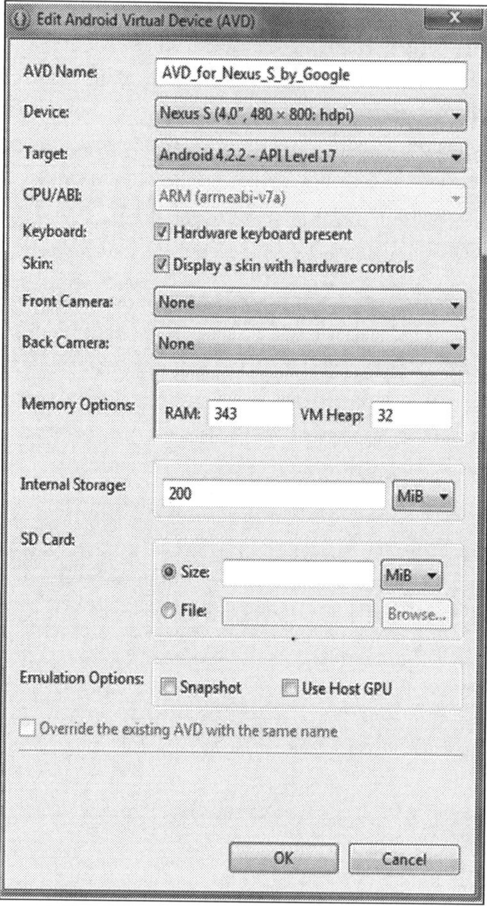

Virtual device configuration

3. Enter the details as per the following information:
    - **AVD Name**: This option is used to provide any name for the virtual device, for example, `ForensicsAVD`.
    - **Device**: This option is used to select any device from the available options based on the screen size.
    - **Target**: This option helps you to select the platform of the device. Note that only the versions that were selected and installed during the SDK installation will be shown here to be selected.
      For our example, the Android 4.4 platform is selected.

- Similarly, you can select hardware features to customize the emulator, for example, the size of internal storage memory, SD card, and so on.

4. A confirmation message is shown once the device is successfully created. Now, select the AVD and click on **Start**. This will prompt you with the launch options. Select any option and click on **Launch**.
5. This should launch the emulator. Note that this could take a few minutes or even longer depending on the workstation's CPU and RAM. The emulator does consume a significant amount of resources on the system. After a successful launch, the AVD will be running as shown in the following screenshot:

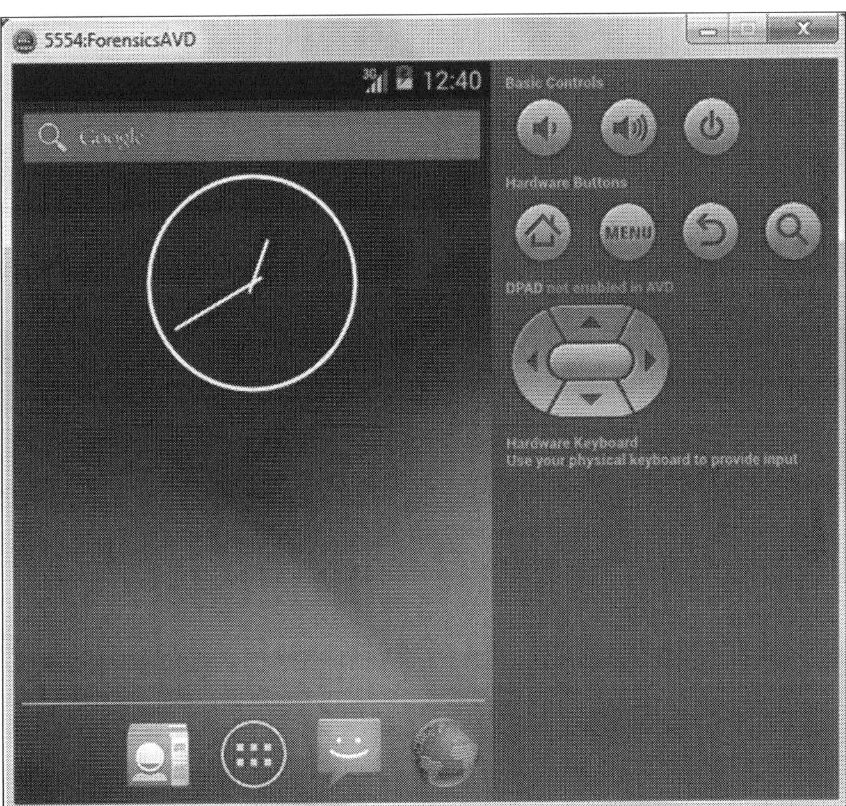

The Android emulator

With emulator, you can configure e-mail accounts, install applications, surf the Internet, send text messages, and more. From a forensic perspective, analysts and security researchers can leverage the functionality of an emulator to understand the file system, data storage, and so on. The data created when working on an emulator is stored in your home directory, in a folder named `.android`. For instance, in our example, the details about the ForensicsAVD emulator that we created earlier are stored under `C:\Users\Rohit\.android\avd\ForensicsAVD.avd`. Among the various files present under this directory, the following are the files that are of interest for a forensic analyst:

- `cache.img`: This is the disk image of the `/cache` partition (remember that we discussed the `/cache` partition of an Android device in *Chapter 7, Understanding Android*).
- `sdcard.img`: This is the disk image of the SD card partition.
- `Userdata-qemu.img`: This is the disk image of the `/data` partition. The `/data` partition contains valuable information about the device user.

Connecting an Android device to a workstation

Forensic acquisition of an Android device using open source tools requires connecting the device to a forensic workstation. Forensic acquisition of any device should be conducted on a forensically sterile workstation. This means that the workstation is strictly used for forensics and not for personal use. Also, note that anytime a device is plugged into a computer, changes can be made to the device. The examiner must have full control of all interactions with the Android device at all times.

The following steps should be performed by the examiner in order to connect the device successfully to a workstation. Note that write protection may prevent the successful acquisition of the device since commands may need to be pushed to the device in order to pull information. All the following steps should be validated on a test device prior to attempting them on real evidence.

Identifying the device cable

The physical USB interface of an Android device allows it to connect to a computer to share data, such as songs, videos, and photos. This USB interface might change from manufacturer to manufacturer and also from device to device. For example, some devices use mini-USB while some others use micro-USB. Apart from this, some manufacturers use their own proprietary formats, such as EXT-USB, EXT micro-USB, and so on. The first step in acquiring an Android device is to determine what kind of device cable is required.

Installing the device drivers

In order to identify the device properly, the computer may need certain drivers to be installed. Without necessary drivers, the computer may not identify and work with the connected device. But the issue is, that since Android is allowed to be modified and customized by the manufacturers, there is no single generic driver that would work for all the Android devices. Each manufacturer writes its own proprietary drivers and distributes them along with the phone. So, it's important to identify specific device drivers, which need to be installed. Of course, some of the Android forensic toolkits (which we are going to discuss in the following chapters) do come with some generic drivers or a set of most-used drivers; they may not work with all the models of Android phones. Some Windows operating systems are able to autodetect and install the drivers once the device is plugged in but more often than not, it fails. The device drivers for each manufacturer can be found on their respective websites.

Accessing the connected device

If you haven't done so already, connect the Android device to the computer directly using the USB cable. The Android device will appear as a new drive and you can access the files on the external storage. Some older Android devices may not be accessible unless the **Turn on USB Storage** option is enabled on the phone as shown in the following screenshot:

USB mass storage

In some Android phones (especially with HTC), the device may expose more than one functionality when connected with a USB cable. For instance, as shown in the following screenshot, when an HTC device is connected, it presents a menu with four options. The default selection is **Charge only**. When the **Disk drive** option is selected, it is mounted as a disk drive.

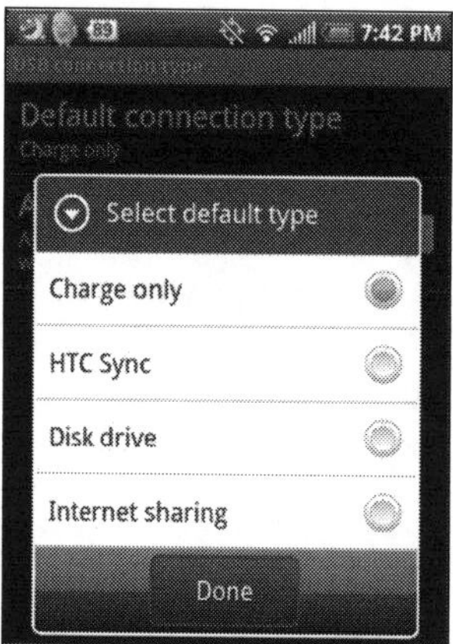

HTC mobile USB options

When the device is mounted as a disk drive, you will be able to access the SD card present on the device. From a forensic point of view, the SD card has significant value as it may contain files that are important for an investigation. However, the core application data stored under `/data/data` will remain on the device and cannot be accessed through these methods.

Android Debug Bridge

Android Debug Bridge (adb) is one of the crucial components in Android forensics. Although we will learn about adb in detail in the coming chapters, we will focus on a basic introduction about adb for now. Android Debug Bridge (adb) is a command-line tool that allows you to communicate with the Android device and control it. You can access the adb tool under `<sdk>/platform-tools/`. Before we discuss anything about adb, we need to have an understanding about the **USB debugging** option. The primary function of this option is to enable communication between the Android device and a workstation on which the Android SDK is installed.

On a Samsung phone, you can access this under **Settings | Developer Options**, as shown in the following screenshot. Other Android phones may have different environments and configuration features. The examiner may have to force the developer options by accessing the build mode. These steps are all device specific and can be determined by researching the device or reading the instructions provided by your forensic tool of choice.

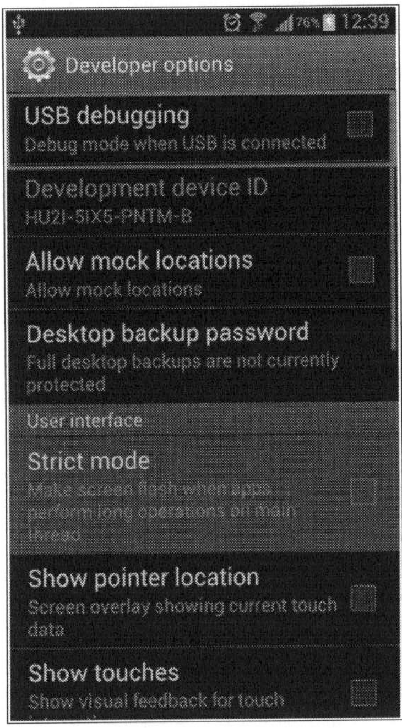

The USB debugging option in Samsung mobiles

When the **USB debugging** option is selected, the device will run adb daemon (`adbd`) in the background and will continuously look for a USB connection. The daemon will usually run under a nonprivileged shell user account and thus will not provide access to complete data. However, on rooted phones, `adbd` will run under the root account and thus provide access to all the data. It is not recommended to root a device to gain full access unless all other forensic methods fail. Should the examiner elect to root an Android device, the methods must be well documented and tested prior to attempting it on real evidence. Rooting will be discussed at the end of this chapter.

On the workstation where the Android SDK is installed, `adbd` will run as a background process. Also, on the same workstation a client program will run, which can be invoked from a shell by issuing the `adb` command. When the adb client is started, it first checks if an adb daemon is already running. If the response is negative, it initiates a new process to start the adb daemon. The adb client program communicates with local `adbd` over port `5037`.

Accessing the device using adb

Once the environment setup is complete and the Android device is in **USB debugging** mode, connect the Android device with the correct USB cable to the forensic workstation and start using adb.

Detecting connected devices

The following adb command provides a list of all the devices connected to the forensic workstation. This would also list the emulator if it is running at the time of issuing the command. Also, remember that if necessary drivers are not installed, then the following command would show a blank message. If you encounter that situation, download the necessary drivers from the manufacturer and install them.

```
C:\android-sdk\platform-tools>adb.exe devices
List of devices attached
4df16ac8115e5f06        device
```

Killing the local adb server

The following command kills the local adb service:

```
C:\android-sdk\platform-tools>adb.exe kill-server
```

After killing the local adb service, issue the adb devices command and observe that the server is started, as shown in the following command lines:

```
C:\android-sdk\platform-tools>adb.exe devices
* daemon not running. starting it now on port 5037 *
* daemon started successfully *
List of devices attached
4df16ac8115e5f06        device
```

Accessing the adb shell

This command allows forensic examiners to access the shell on an Android device and interact with the device. The following is the command to access the adb shell and execute a basic `ls` command to see the contents of the current directory:

```
C:\android-sdk\platform-tools>adb.exe shell
shell@android:/ $ ls
ls
acct
cache
config
```

```
d
data
default.prop
dev
efs
etc
factory
fstab.smdk4x12
init
init.bt.rc
init.goldfish.rc
init.rc
init.smdk4x12.rc
init.smdk4x12.usb.rc
....
```

The Android emulator can be used by forensic examiners to execute and understand adb commands before using them on the device. In *Chapter 9, Android Data Extraction Techniques*, we are going to explain more about leveraging adb to install applications, copy files and folders from the device, view device logs, and so on.

Handling an Android device

Handling an Android device in a proper manner prior to the forensic investigation is a very important task. Care should be taken to make sure that our unintentional actions don't result in data modification or any other unwanted happenings. The following sections throw light on certain issues which need to be considered while handling the device in the initial stages of forensic investigation.

With the improvements in technology, the concept of **device locking** has effectively changed over the last few years. Most users now have a passcode locking mechanism enabled on their device due to the increase in general security awareness. Before we look at some of the techniques to bypass the locked Android devices, it is important not to miss an opportunity to disable the passcode when there is a chance.

When an Android device, which is to be analyzed, is first accessed, check if the device is still active (unlocked). If so, change the settings of the device to enable greater access to the device. So, when the device is still active, consider performing the following tasks:

- **Enabling USB debugging**: Once the USB debugging option is enabled, it gives greater access to the device through the adb connection. This is of great significance when it comes to extracting data from the device. The location to enable USB debugging might change from device to device but it's usually under **Developer Options** in **Settings**. Most methods for physically acquiring Android devices require USB debugging to be enabled.
- **Enabling the "Stay awake" setting**: If the **Stay awake** option is selected and the device is connected for charging, then the device never locks. Again, if the device locks, the acquisition could be halted.
- **Increasing screen timeout**: This is the time for which the device will be effectively active once it is unlocked. The location to access this setting varies depending upon the model of the device. On a Samsung Galaxy S3 phone, you can access the same under **Settings | Display | Screen Timeout**.

Apart from this, as mentioned in *Chapter 1, Introduction to Mobile Forensics*, the device needs to be isolated from the network to make sure that remote wipe options do not work on the device. The Android Device Manager allows the phone to be remotely wiped or locked. This can be done by signing in to the Google account, which is configured on the mobile. More details about this are mentioned in the following section. If the Android device is not set up to allow remote wiping, the device can only be locked using the Android Device Manager. Also, there are several **Mobile Device Management (MDM)** software products available on the market, which allow users to remotely lock or wipe the Android device. Some of these may not require specific settings to be enabled on the device.

Using the available remote wipe software, it is possible to delete all the data including e-mails, applications, photos, contacts, and other files including those found on the SD card. To isolate the device from the network, you can put the device in airplane mode and disable Wi-Fi as an extra precaution. Enabling airplane mode and disabling Wi-Fi works well as the device will not be able to communicate over a cellular network and cannot be accessed via Wi-Fi. Removing the SIM card from the phone is also an option but that does not effectively stop the device from communicating over Wi-Fi or some cellular networks. To place the device in airplane mode, press and hold the **Power/Off** button and select airplane mode.

All these steps can be done when the Android device is not locked. However, during the investigation, we commonly encounter devices that are locked. Hence, it's important to understand how to bypass the lock code if it is enabled on an Android device.

Screen lock bypassing techniques

Due to the increase in user awareness and the ease of functionality, there has been an exponential increase in the usage of passcode options to lock Android devices. Hence, bypassing the device's screen lock during a forensic investigation is becoming increasingly important. The screen lock bypass techniques discussed have their applicability based on the situation. Note that some of these methods are used to make changes to the device. Make sure that you test and validate all the steps listed on non-evidentiary Android devices. The examiner must have authorization to make the required changes to the device, document all steps taken, and be able to describe the steps taken if a courtroom testimony is required.

Currently, there are three types of screen lock mechanisms offered by Android. Although there are some devices which have voice lock and face lock options, we will limit our discussion to the following three options since these are most widely used on all Android devices:

- **Pattern Lock**: The user sets a pattern or design on the phone and the same must be drawn to unlock the device. Android was the first smartphone to introduce a pattern lock.
- **PIN code**: This is the most common lock option and is found on many mobile phones. The PIN code is a 4-digit number that needs to be entered to unlock the device.
- **Passcode (alphanumeric)**: This is an alphanumeric passcode. Unlike the PIN, which takes four digits, the alphanumeric passcode takes more than just digits.

The following section details some of the techniques to bypass these Android lock mechanisms. Depending on the situation, these techniques might help an investigator to bypass the screen lock.

Using adb to bypass the screen lock

If USB debugging appears to be enabled on the Android device, it is wise to take advantage of it by connecting with adb using USB, as discussed in the earlier sections. The examiner should connect the device to the forensic workstation and issue the adb devices command. If the device shows up, it implies that USB debugging is enabled. If the Android device is locked, the examiner must attempt to bypass the screen lock. The following are the two methods that may allow the examiner to bypass the screen lock when USB debugging is enabled.

Deleting the gesture.key file

This is how the process is done:

1. Connect the device to the forensic workstation (a Windows machine in our example) using a USB cable.
2. Open the command prompt and execute the following instructions:

```
adb.exe shell
cd /data/system
rm gesture.key
```

3. Reboot the device. If the pattern lock still appears, just draw any random design and observe that the device should unlock without any trouble.

   This method works when the device is rooted. This method may not be successful on unrooted devices. Rooting an Android device should not be performed without proper authorization as the device is altered.

Updating the settings.db file

To update the `settings.db` file, perform the following steps:

1. Connect the device to the forensic workstation using a USB cable.
2. Open the command prompt and execute the following instructions:

```
adb.exe shell
cd /data/data/com.android.providers.settings/databases
sqlite settings.db
sqlite>update system set value=0 where
name='lock_pattern_autolock';
sqlite>update system set value=0 where name=
'lockscreen.lockedoutpermenantly';
```

3. Exit and reboot the device.
4. The Android device should be unlocked. If not, attempt to remove `gesture.key` as explained earlier.

Checking for the modified recovery mode and adb connection

In Android, recovery refers to the dedicated partition where the recovery console is present. The two main functions of recovery are to delete all user data and install updates. For instance, when you factory reset your phone, recovery boots up and deletes all the data. Similarly, when updates are to be installed on the phone, it is done in the recovery mode. There are many enthusiastic Android users who install custom ROM through a modified recovery module. This modified recovery module is mainly used to make the process of installing custom ROM easy. Recovery mode can be accessed in different ways depending on the manufacturer of the device, which is easily available on the Internet. Usually, this is done by holding different keys together such as the volume button and power button. Once in recovery mode, connect the device to the workstation and try to access the adb connection. If the device has a recovery mode which is not modified, the examiner may not be able to access the adb connection. The modified recovery versions of the device present the user with different options and can be easily noticed.

Flashing a new recovery partition

There are mechanisms available to flash the recovery partition of an Android device with a modified image. The **Fastboot** utility would facilitate this process. Fastboot is a diagnostic protocol that comes with the SDK package, used primarily to modify the flash file system through a USB connection from a host computer. For this, you need to start the device in the boot loader mode in which only the most basic hardware initialization is performed. Once the protocol is enabled on the device, it will accept a specific set of commands that are sent to it via the USB cable using a command line. Flashing or rewriting a partition with a binary image stored on the computer is one such command that is allowed. Once the recovery is flashed, boot the device in recovery mode, mount the `/data` and `/system` partitions, and use adb to remove the `gesture.key` file. Reboot the phone and you should be able to bypass the screen lock.

Smudge attack

In rare cases, a smudge attack may be used to deduce the password of a touchscreen mobile device. The attack relies on identifying the smudges left behind by the user's fingers. While this may present a bypass method, it must be said that a smudge attack is unlikely since most Android devices are touchscreen and smudges will also be present from using the device. However, it has been demonstrated that under proper lighting, the smudges that are left behind can be easily detected as shown in the following screenshot (`http://www.securitylearn.net/tag/android-passcode-bypass/`). By analyzing the smudge marks, we can discern the pattern that is used to unlock the screen. This attack is more likely to work while discerning the pattern lock on the Android device. In some cases, PIN codes can also be recovered depending upon the cleanliness of the screen. So, during a forensic investigation, care should be taken when the device is first handled to make sure that the screen is not touched.

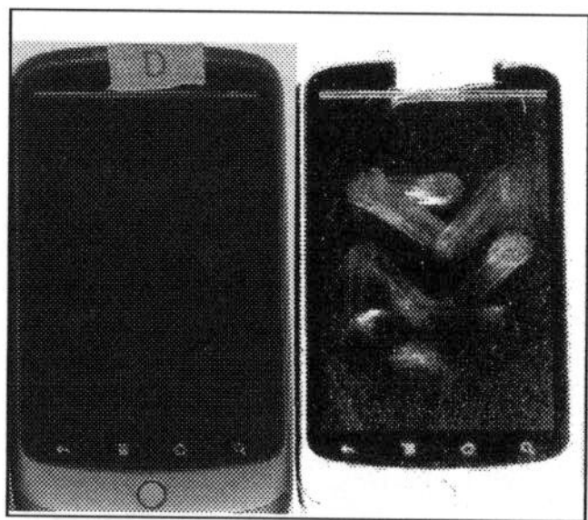

Smudges visible on a device under proper lighting (source: https://viaforensics.com/wpinstall/wp-content/uploads/smudge.png)

Using the primary Gmail account

If you know the username and password of the primary Gmail address that is configured on the device, you can change the PIN, password, or swipe on the device. After making a certain number of failed attempts to unlock the screen, Android provides an option named **Forgot Pattern** or **Forgot Password** as shown in the following screenshot. Tap on that link and sign in using the Gmail username and password and this will allow you to create a new pattern lock or passcode for the device.

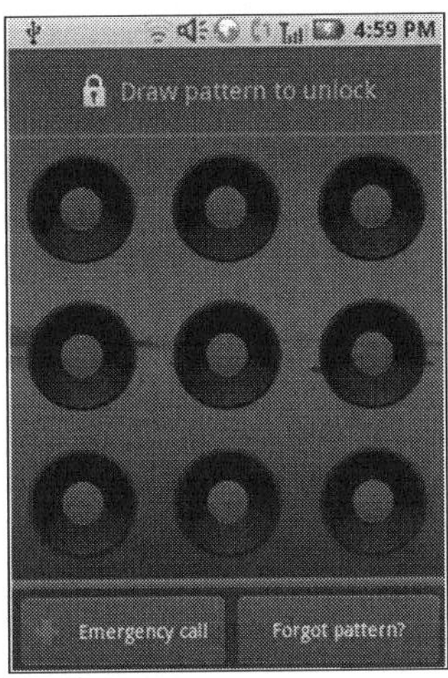

Forgot pattern option on an Android device

Other techniques

All of the earlier mentioned techniques and the commercial tools available prove to be useful to the forensic examiner trying to get access to the data on the Android devices. However, there could be situations where none of these techniques work. To obtain a complete physical image of the device, techniques such as chip-off and JTAG may be required when commercial and open source solutions fail. A short description of these techniques is mentioned.

While the chip-off technique removes the memory chip from a circuit and tries to read it, the JTAG technique involves probing the JTAG **Test Access Ports (TAPs)** and soldering connectors to the JTAG ports in order to read data from the device memory. The chip-off technique is more destructive because once the chip is removed from the device, it is difficult to restore the device back to its original functional state. Also, expertise is needed to carefully remove the chip from the device by desoldering the chip from the circuit board. The heat required to remove the chip can also damage or destroy the data stored on that chip. Hence, this technique should be looked upon only when the data is not retrievable by open source or commercial tools or the device is damaged beyond repair. When using the JTAG technique, JTAG ports help an examiner to access the memory chip to retrieve a physical image of the data without needing to remove the chip. To turn off the screen lock on a device, an examiner can identify where the lock code is stored in the physical memory dump, turn off the locking, and copy that data back to the device. Commercial tools, such as Cellebrite Physical Analyzer, can accept `.bin` files from chip-off and JTAG acquisitions and crack the lock code for the examiner. Once the code is either manually removed or cracked, the examiner can analyze the device using normal techniques.

Both the chip-off and JTAG techniques require extensive research and experience to be tried on a real device. A great resource for JTAG and chip-off on devices can be found at `http://www.forensicswiki.org/wiki`.

Gaining root access

As a mobile device forensic examiner, it is essential to know everything that relates to twisting and tweaking the device. This would help you to understand the internal working of the device in detail and comprehend many issues that you may face during your investigation. Rooting Android phones has become a common phenomenon and you can expect to encounter rooted phones during forensic examinations. The examiner, where applicable, may also need to root the device in order to acquire data for the forensic examination. Hence, it's important to know the ins and outs of rooted devices and how they are different from the other phones. The following sections cover information about Android rooting and other related concepts.

What is rooting?

The default administrative account in Unix-like operating systems is called "root". So, in Linux, the root user has the power to start/stop any system service, edit/delete any file, change the privileges of other users, and so on. We have already learned that Android uses the Linux kernel and hence most of the concepts present in Linux are applicable to Android as well. However, when you buy an Android phone, it does not let you log in as a root user by default. Rooting an Android phone is all about gaining access on the device to perform actions that are not normally allowed on the device. Manufacturers want the devices to function in a certain manner for normal users. Rooting a device may void a warranty since root opens the system to vulnerabilities and provides the user with **superuser** capabilities. Imagine a malicious application having access to an entire Android system with root access. Remember that in Android, each application is treated as a separate user and issues a UID. Thus, the applications have access to limited resources and the concept of application isolation is enforced. Essentially, rooting an Android device allows superuser capabilities and provides open access to the Android device.

Rooting an Android device

Even though the hardware manufacturers try to put enough restrictions to restrict access to the root, hackers have always found different ways to get access to the root. The process of rooting varies depending on the underlying device manufacturer. But rooting any device usually involves exploiting a security bug in the device's firmware and then copying the `su` (superuser) binary to a location in the current process's path (`/system/xbin/su`) and granting it executable permissions with the `chmod` command.

For the sake of simplicity, imagine that an Android device has three to four partitions, which run programs not entirely related to Android (Android being one among them).

The boot loader is present in the first partition and is the first program that runs when the phone is powered on. The primary job of this boot loader is to boot other partitions and load the Android partition, commonly referred to as ROM by default. To see the boot loader menu, a specific key combination is required such as holding the power button and pressing the volume up button. This menu provides options for you to boot into other partitions such as the recovery partition.

The recovery partition deals with installing upgrades to the phone, which are written directly to the Android ROM partition. This is the mode that you see when you install any official update on the device. Device manufacturers make sure that only official updates are installed through the recovery partition. Thus, bypassing this restriction would allow you to install/flash any unlocked Android ROM. Modified recovery programs are those that not only allow an easier rooting process but also provide various options, which are not seen in the normal recovery mode. The following screenshot shows the normal recovery mode:

```
Android system recovery <3e>

Volume up/down to move highlight:
power button to select.

reboot system now
apply update from ADB
update/recover from SD card
wipe data/factory reset
wipe cache partition
```

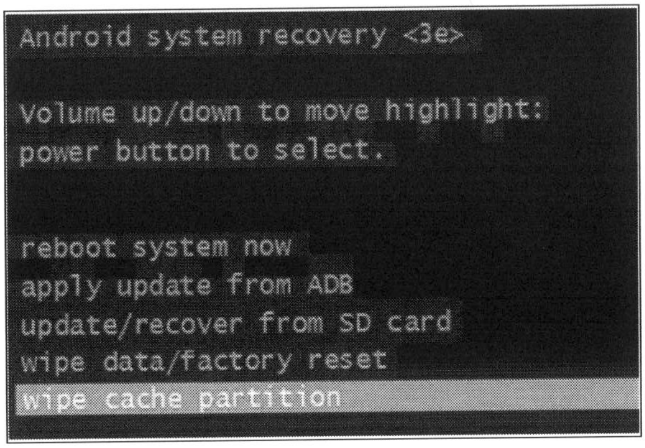

Normal Android system recovery mode

The following screenshot shows the modified recovery mode:

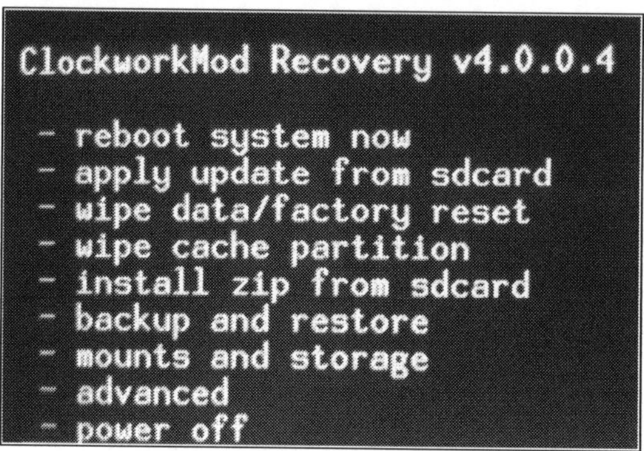

Modified recovery mode

The most used recovery program in the Android world is the **Clockwork** recovery, also called **ClockworkMod**. Hence, most of the rooting methods begin by flashing a modified recovery to the recovery partition. After that, you can issue an update, which can root the device. However, you don't need to perform all the actions manually as software is available for most of the models, which could root your phone with a single click.

Rooting a device has both advantages and disadvantages associated with it.

The following are the advantages:

- Rooting allows modification of the software on the device to the deepest level. For example, you can overclock or underclock the device's CPU (`http://techbeasts.com/2014/01/17/what-is-cpu-underclocking-overclocking-and-how-to-underclock-overclock/`).
- Bypass restrictions imposed on the device by carriers, manufacturers, and so on.
- For extreme customization, new customized ROMs could be downloaded and installed.

The following are the disadvantages:

- Rooting a device must be done with extreme care as errors may result in irreparable damage to the software on the phone turning the device into a useless brick.
- Rooting might void the warranty of a device.
- Rooting results in increased exposure to malware and other attacks. Malware with access to the entire Android system can create havoc.

Once the device is rooted, applications such as the Superuser app are available to provide and deny root privileges. This app helps you to grant and manage superuser rights on the device, as shown in the following screenshot:

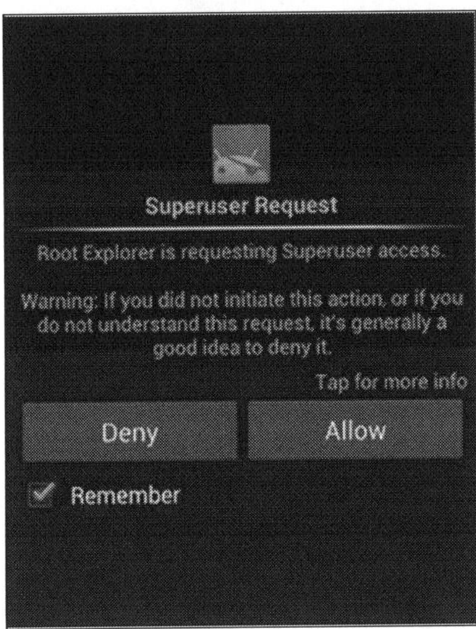

Application requesting root access

Root access – adb shell

A normal Android phone does not allow you to access certain directories and files on the device. For example, try to access the `/data/data` folder on an Android device, which is not rooted. You will see the following message:

```
C:\android-sdk\platform-tools>adb.exe shell
shell@android:/ $ cd /data/data
cd /data/data
shell@android:/data/data $ ls
ls
opendir failed, Permission denied
255|shell@android:/data/data $
```

On a rooted phone, you can run the adb shell as a root by issuing the following command:

```
C:\android-sdk\platform-tools>adb.exe root
restarting adbd as root
```

```
C:\android-sdk\platform-tools>adb.exe shell
root@android:/ # cd /data/data
cd /data/data
root@android:/data/data # ls
ls
com.adobe.flashplayer
com.adobe.reader
com.aldiko.android
com.android.backupconfirm
com.android.browser
```

Thus, rooting a phone enables you to access folders and data, which are otherwise not accessible. Also, note that # symbolizes root or superuser access while $ reflects a normal user, as shown in the preceding command lines.

Summary

A proper forensic workstation setup is required prior to conducting investigations on an Android device. Using open source methods to acquire and analyze Android devices requires the installation of specific software on the forensic workstation. If the method of forensic acquisition requires the Android device to be unlocked, the examiner needs to determine the best method to gain access to the device. Various screen lock bypass techniques explained in this chapter help an examiner to bypass the passcode under different circumstances. Depending on the forensic acquisition method and scope of the investigation, rooting the device should provide complete access to the files present on the device. Some commercial tools, such as Micro Systemation XRY, provide a root that the examiner must use in order to access specific areas of the device memory. Now that the basic concepts are covered on gaining access to an Android device, we will cover acquisition techniques and describe how the data is being pulled using each method in *Chapter 9, Android Data Extraction Techniques*.

9
Android Data Extraction Techniques

By using any of the passcode bypass techniques explained in *Chapter 8, Android Forensic Setup and Pre Data Extraction Techniques*, an examiner can try to access a locked device. Once the device is accessible, the next task is to extract the information present on the device. This can be achieved by applying various data extraction techniques on the Android device. This chapter helps you to identify the sensitive locations present on an Android device and explains various logical and physical techniques that can be applied to the device in order to extract the necessary information.

Imaging an Android Phone

Imaging a device is one of the most important steps in mobile device forensics. The rule of thumb when dealing with a forensic examination is to ensure that the data present on the device is not modified in any way, wherever possible. As explained in *Chapter 1, Introduction to Mobile Forensics*, all the changes by the examiner from the previous testing and validation should be well documented. When possible, it's imperative to obtain a physical image of the Android device before performing any techniques to extract the data directly from the device. In forensics, this process of obtaining a physical or logical acquisition is commonly called **imaging the device**. A physical image is preferred as it is a bit-by-bit copy of the Android device memory.

It is important to understand that a bit-by-bit image is not similar to copying and pasting the contents on the device. If we copy and paste the contents on a device it will only copy the available files such as visible files, hidden files, and system-related files. This method is considered a logical image. With this method, deleted files and files that are not accessible are not copied by the copy command. Deleted files can be recovered (based on the circumstances) using certain techniques, which we are going to see in the following chapters. Hence, you need to take a 1:1 bit-by-bit image of the device memory to obtain all of the data.

Let's first revisit how imaging is done on a desktop computer as it helps us to correlate and realize the problems associated with imaging Android devices. Let's assume that a desktop computer, which is not powered on, is seized from a suspect and sent for forensic examination. In this case, a typical forensic examiner would remove the hard disk, connect it to a write blocker and obtain a bit-by-bit forensic image using any of the available tools. The original hard disk is then safely protected during the forensic imaging of the data. With an Android device, all the areas that contain data cannot be easily removed. Also, if the device is active at the time of receiving it for examination, it is not possible to analyze the device without making any changes to it because any interaction would change the state of the device.

An Android device may have two file storage areas, internal and external storage. Internal storage refers to the built-in non-volatile memory. External storage refers to the removable storage medium such as a micro SD card. However, it's important to note that some devices do not have a removable storage medium such as an SD card, but they divide the available permanent storage space into internal and external storage. Hence, it's not always true that external storage is something that is removable. When a removable SD card is present, a forensic image of the memory card has to be obtained. As discussed in *Chapter 7, Understanding Android*, these removable cards are generally formatted with the FAT 32 file system. Some mobile device acquisition methods will acquire the SD card through the Android device. This process, while useful, will be slow due to the speed limitations of the USB phone cables.

Data extraction techniques

Data residing on an Android device may be an integral part of civil, criminal, or internal investigations done as part of a corporate company's internal probe. While dealing with investigations involving Android devices, the forensic examiner needs to be mindful of the issues that need to be taken care of during the forensic process; this includes determining if root access is permitted (via consent or legal authority) and what data can be extracted and analyzed during the investigation. For example, in a criminal case involving stalking, the court may only allow for the SMS, call logs, and photos to be extracted and analyzed on the Android device belonging to the suspect. In this case, it may make the most sense to logically capture just those specific items. However, it is best to obtain a full physical data extraction of the device and only examine the areas admissible by the court. You never know where your investigation may lead and it is best to obtain as much data off the device immediately rather than wish you had a full image should the scope of consent change.

The data extraction techniques on an Android device can be classified into three types:

- Manual data extraction
- Logical data extraction
- Physical data extraction

The extraction methods for each of these types will be described in detail in the following sections. Some methods may require the device be rooted in order to fully access the data. Each method has different implications and success rates will depend on the tool, method used, and device make and model.

Manual data extraction

This method of extraction involves the examiner utilizing the normal user interface of the mobile device to access content present in the memory. The examiner will browse through the device normally by accessing different menus to view the details such as call logs, text messages, and IM chats. The content of each screen is captured by taking pictures and can be presented as evidence. The main drawback with this type of examination is that only those files that are accessible by the operating system (in the UI mode) can be investigated. Care must be taken when manually examining the device as it's easy to press the wrong button and erase or add data. Manual extraction should be used as a last resort to verify findings extracted using one of the other methods. Certain circumstances may warrant the examiner to conduct manual examination as the first step. This may include life or death situations or missing persons where a quick scan of the device may lead the police to the individual.

Using root access to acquire an Android device

Android, by default, does not provide access to the internal directories and system-related files. This restricted access is to ensure the security of the device. For instance, the `/data/data` folder is not accessible on a non-rooted device. This folder is especially of interest to us because it stores most of the user-created data and many applications write valuable data into this folder. Hence, to obtain an image of the device, we need to root the Android device. Rooting a device gives us the **superuser** privileges and access to all the data. It is important to realize that this book has been stressing that all the steps taken should be forensically sound and not make changes to the device whenever possible. Rooting an Android device will make changes to it and should be tested on any device that the examiner has not previously investigated. Rooting is common for Android devices, but getting root access could alter the device in a manner that renders the data changed or worse yet—wiped. Some Android devices, such as the Nexus 4 and 5, may force the data partition to be wiped prior to allowing root access. This negates the need to root the device in order to gain access because all the user data is lost during the process. Just remember that while rooting provides access to more data when successfully done, it can also wipe the data or destroy the phone. Hence, you must ensure you have consent or legal rights to manipulate the Android device prior to proceeding with the root. As rooting techniques have been discussed in *Chapter 8, Android Forensic Setup and Pre Data Extraction Techniques*, we will proceed with the example assuming that the device is rooted. The following is a step-by-step process to obtain a forensic image of a rooted Android device.

Install the **Android Terminal Emulator** application. The Android Terminal Emulator application helps you to access the Linux command shell. Android Terminal Emulator can be downloaded from `https://github.com/jackpal/Android-Terminal-Emulator/wiki`. Once installed, you can run most of the Linux commands on the device. It is recommended to install it through adb instead of connecting to the Internet to install it from the Google Play store. The following screenshot shows the installation of the Android Terminal Emulator application on a Mac running v10.9.2:

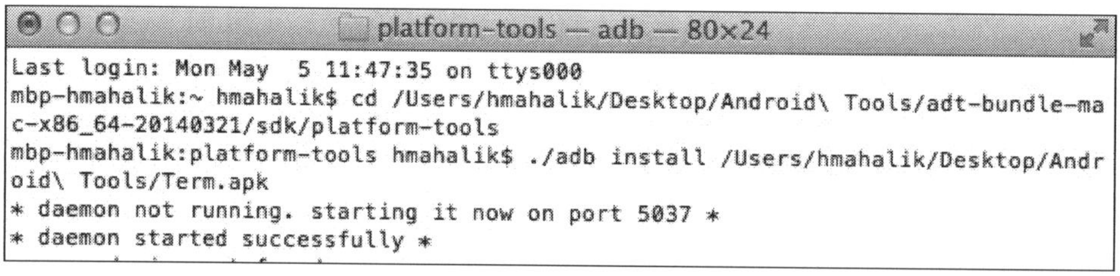

```
platform-tools — adb — 80×24
Last login: Mon May  5 11:47:35 on ttys000
mbp-hmahalik:~ hmahalik$ cd /Users/hmahalik/Desktop/Android\ Tools/adt-bundle-ma
c-x86_64-20140321/sdk/platform-tools
mbp-hmahalik:platform-tools hmahalik$ ./adb install /Users/hmahalik/Desktop/Andr
oid\ Tools/Term.apk
* daemon not running. starting it now on port 5037 *
* daemon started successfully *
```

Once Android Terminal Emulator is installed, the partitions can be acquired from the Android device using the following steps:

- **Using the dd command**: The `dd` command can be used to create a raw image of the device. This command helps us to create a bit-by-bit image of the Android device by copying low-level data.
- **Inserting a new SD card**: Insert a new SD card into the device in order to copy the image file to this card. Make sure this SD card is wiped and does not contain any other data.
- **Executing the command**: The file system of an Android device is stored in different locations within the `/dev` partition. A simple mount command on a Samsung Galaxy S3 phone returns the following output:

```
shell@Android:/ $ mount
mount
rootfs / rootfs ro,relatime 0 0
tmpfs /dev tmpfs rw,nosuid,relatime,mode=755 0 0
devpts /dev/pts devpts rw,relatime,mode=600 0 0
proc /proc proc rw,relatime 0 0
sysfs /sys sysfs rw,relatime 0 0
none /acct cgroup rw,relatime,cpuacct 0 0
tmpfs /mnt/asec tmpfs rw,relatime,mode=755,gid=1000 0 0
tmpfs /mnt/obb tmpfs rw,relatime,mode=755,gid=1000 0 0
none /dev/cpuctl cgroup rw,relatime,cpu 0 0
/dev/block/mmcblk0p9 /system ext4 ro,noatime,barrier=1,data=order
ed 0 0
/dev/block/mmcblk0p3 /efs ext4 rw,nosuid,nodev,noatime,barrier=1,j
ournal_async_c
ommit,data=ordered 0 0
/dev/block/mmcblk0p8 /cache ext4 rw,nosuid,nodev,noatime,barrier=1
,journal_async
_commit,data=ordered 0 0
/dev/block/mmcblk0p12 /data ext4 rw,nosuid,nodev,noatime,barrier=1
,journal_async
_commit,data=ordered,noauto_da_alloc,discard 0 0
/sys/kernel/debug /sys/kernel/debug debugfs rw,relatime 0 0
/dev/fuse /storage/sdcard0 fuse rw,nosuid,nodev,noexec,relatime,us
er_id=1023,group_id=1023,def
ault_permissions,allow_other 0 0
```

From the preceding output, we can identify the blocks where the `/system`, `/data`, and `/cache` partitions are mounted. Although it's important to image all the files, most of the data is present in the `/data` and `/system` partitions. When time allows, all partitions should be acquired for completeness. Once this is done, execute the following command to image the device:

```
dd if=/dev/block/mmcblk0p12 of=/sdcard/tmp.image
```

In the preceding example, the data partition of a Samsung Galaxy SIII was used (where `if` is the input file and `of` is the output file).

The preceding command will make a bit-by-bit image of the `mmcblk0p12` file (data partition) and copy the image file to an SD card. Once this is done, the `dd` image file can be analyzed using the available forensic software.

The examiner must ensure that the SD card has enough storage space to contain the data partition image. Other methods are available to acquire data from the rooted devices.

Logical data extraction

Logical data extraction techniques extract the data present on the device by accessing the file system. These techniques are significant because they provide valuable data, work on most devices, and are easy to use. Once again, the concept of rooting comes into picture while extracting the data. Logical techniques do not actually require root access for data extraction. However, having root access on a device allows you to access all the files present on a device. This means that some data may be extracted on a non-rooted device while root access will open the device and provide access to all the files present on the device. Hence, having root access on a device would greatly influence the amount and kind of data that can be extracted through logical techniques. Logical extraction can be performed on a device in two ways:

- Using adb pull commands
- Using content providers

The following sections explain each of these options and how the data can be extracted.

Using the adb pull command

As seen earlier, adb is a command-line tool that helps you communicate with the device to retrieve information. Using adb, you can extract data from all the files on the device or only the relevant files in which you are interested. To access an Android device through adb, it's necessary that the USB debugging option is enabled. If the device is locked and USB debugging is not enabled, try to bypass the screen lock using the techniques mentioned in *Chapter 8, Android Forensic Setup and Pre Data Extraction Techniques*.

As a forensic examiner, it's important to know how the data is stored on the Android device and to understand where important and sensitive information is stored so that the data can be extracted accordingly. Application data often contains a wealth of user data that may be relevant to the investigation. All files pertaining to applications of interest should be examined for relevance, as will be explained in *Chapter 10, Android Data Recovery Techniques*. The application data can be stored in one of the following locations:

- **Shared preferences**: Data is stored in key-value pairs in a lightweight XML format. Shared preference files are stored in the `shared_pref` folder of the application `/data` directory.
- **Internal storage**: Data stored here is private and is present in the device's internal memory. Files saved to the internal storage are private and cannot be accessed by other applications.
- **External storage**: This stores data that is public in the device's external memory, which does not usually enforce security mechanisms. This data is available under the `/sdcard` directory.
- **SQLite database**: This data is available in the `/data/data/PackageName/` database. They are usually stored with a `.db` file extension. The data present in a SQLite file can be viewed using a SQLite browser (`http://sourceforge.net/projects/SQLitebrowser/`) or by executing the necessary SQLite commands on the respective files.

Every Android application stores the data on the device using any of the preceding data storage options. So, the **Contacts** application would store all the information about the contact details in the `/data/data` folder under its package name. Note that `/data/data` is a part of your device's internal storage where all the apps are installed under normal circumstances. Some application data will reside on the SD card and in the `/data/data` partition. Using adb, we can pull the data present in this partition for further analysis using the adb pull command. Once again, it's important to note that this directory is accessible only on a rooted phone.

Extracting the /data directory on a rooted device

On a rooted phone, a pull command on `/data` can be executed as follows:

```
C:\android-sdk-windows\platform-tools>adb.exe pull /data C:\temp
pull:
/data/data/com.kiloo.subwaysurf/app_sslcache/www.chartboost.com.443 -
>
C:\temp/data/com.kiloo.subwaysurf/app_sslcache/www.chartboost.com.443
pull: /data/data/com.mymobiler.android/lib/libpng2.so -> C:\temp/data/
com.mymobiler.android/lib/libpng2.so


pull: /data/system.notfirstrun -> C:\temp/system.notfirstrun
732 files pulled. 0 files skipped.
2436 KB/s (242711369 bytes in 97.267s)
```

As shown in the following screenshot, the complete `/data` directory on the Android device was copied to the local directory on the machine. The entire data directory was extracted in 97 seconds. The extraction time will vary depending on the amount of data residing in `/data`.

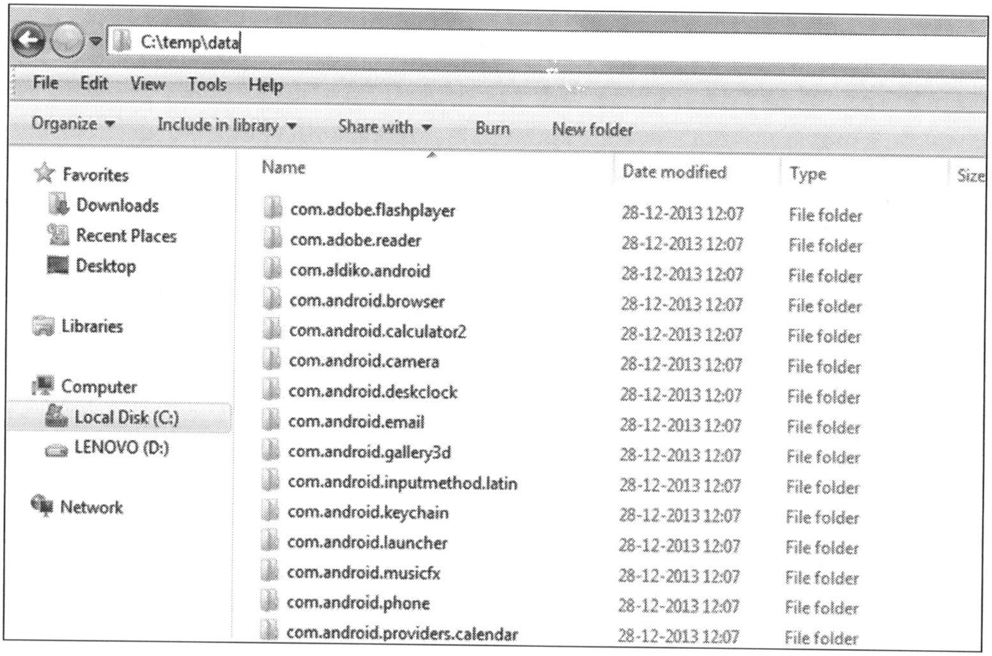

The/data directory extracted to a forensic workstation

On a non-rooted device, a pull command on the /data directory does not extract the files as shown in the following output, since the shell user does not have permission to access those files:

```
C:\android-sdk-windows\platform-tools>adb.exe pull /data C:\temp
pull: building file list...
0 files pulled. 0 files skipped.
```

The data copied from a rooted phone through the preceding process maintains the directory structure, thus allowing an investigator to browse through the necessary files to gain access to the information. By analyzing the data of the respective applications, a forensic expert can gather critical information that can influence the outcome of the investigation. Note that examining the folders natively on your forensic workstation will alter the dates and times of the content. The examiner should make a copy of the original output to use for a date/time comparison.

Using SQLite Browser

SQL Browser is a tool that can help during the course of analyzing the extracted data. SQLite Browser allows you to explore the database files with the following extensions: .sqlite, .sqlite3, .sqlitedb, .db, and .db3. The main advantage of using SQLite Browser is that it shows the data in a table form. Navigate to **File** | **Open Database** to open a .db file using SQLite Browser. As shown in the following screenshot, there are three tabs: **Database Structure**, **Browse Data**, and **Execute SQL**. The **Browse Data** tab allows you to see the information present in different tables within the .db files. We will be mostly using this tab during our analysis. Alternately, **Oxygen Forensic SQLite Database Viewer** can also be used for the same purpose. Recovering deleted data from database files is possible and will be explained in *Chapter 10, Android Data Recovery Techniques.*

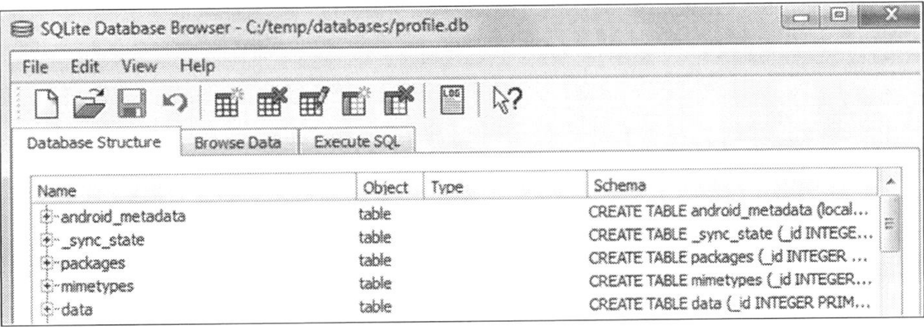

SQLite Browser

The following sections throw light on identifying important data and manually extracting various details from an Android phone.

Extracting device information

Knowing the details of your Android device, such as the model, version, and more, will aid in your investigation. For example, when the device is physically damaged and prohibits the examination of the device information, you can grab the details about the device by viewing the `build.prop` file present in the `/system` folder, as follows:

```
shell@android:/system $ cat build.prop
cat build.prop
# begin build properties
# autogenerated by buildinfo.sh
ro.build.id=JZO54K
ro.build.display.id=JZO54K.I9300XXEMH4
ro.build.version.incremental=I9300XXEMH4
ro.build.version.sdk=16
ro.build.version.codename=REL
ro.build.version.release=4.1.2
ro.build.date=Tue Sep 17 17:26:31 KST 2013
ro.build.date.utc=1379406391
.. .. ..
ro.product.model=GT-I9300
ro.product.brand=samsung
ro.product.name=m0xx
ro.product.device=m0
ro.product.board=smdk4x12
ro.product.cpu.abi=armeabi-v7a
ro.product.cpu.abi2=armeabi
ro.product_ship=true
ro.product.manufacturer=samsung
.. .. ..
ro.build.description=m0xx-user 4.1.2 JZO54K I9300XXEMH4 rel
ro.build.fingerprint=samsung/m0xx/m0:4.1.2/JZO54K/I9300XXEM
.. .. ..
ro.build.PDA=I9300XXEMH4
```

```
ro.build.hidden_ver=I9300XXEMH4
.. .. ..
ro.sec.fle.encryption=true
.. .. ..
ro.com.google.gmsversion=4.1_r6
dalvik.vm.dexopt-flags=m=y
net.bt.name=Android
dalvik.vm.stack-trace-file=/data/anr/traces.txt
```

Extracting call logs

Accessing the call logs of a phone is often required during the investigation to confirm certain events. The information about call logs is stored in the `contacts2.db` file located at `/data/data/com.android.providers.contacts/databases/`. As mentioned earlier, you can use SQLite Browser to see the data present in this file after extracting it to a local folder on the forensic workstation. As shown in the following screenshot, by using the adb pull command, the necessary `.db` files can be extracted to a folder on the forensic workstation, as shown in the following screenshot:

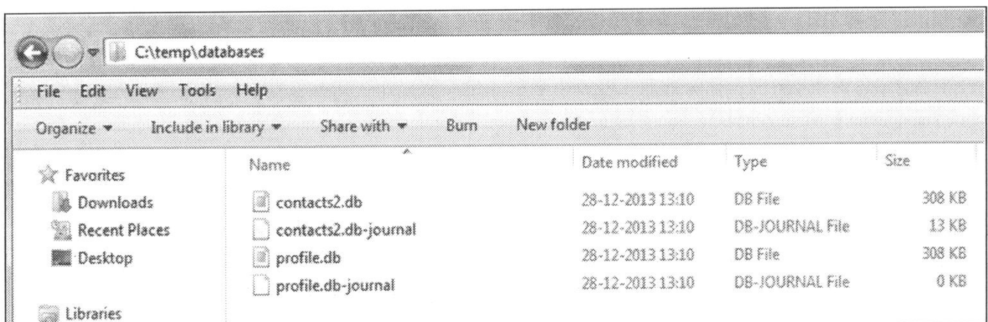

The contacts2.db file copied to a local folder

Note that applications used to make calls can store call log details in the respective application folder. All communication applications must be examined for call log details, as follows:

```
C:\android-sdk-windows\platform-tools>adb.exe pull
/data/data/com.android.providers.contacts C:\temp
pull: building file list...
.. .. .. ..
pull:
/data/data/com.android.providers.contacts/databases/contacts2.db ->
C:\temp/databases/contacts2.db
```

```
pull: /data/data/com.android.providers.contacts/databases/profile.db
-> C:\temp/databases/profile.db

pull: /data/data/com.android.providers.contacts/databases/profile.db-
journal ->C:\temp/databases/profile.db-journal

6 files pulled. 0 files skipped.

70 KB/s (644163 bytes in 8.946s)
```

Now, open the `contacts2.db` file using SQLite Browser (navigating to **File | Open Database**) and browse through the data present in different tables. The `calls` table present in the `contacts2.db` file provides information about the call history. The following screenshot highlights the call history along with the **name**, **number**, **duration**, and **date**.

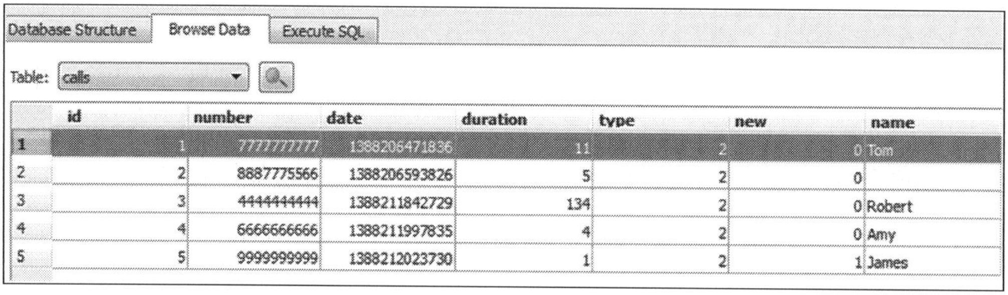

Database Structure | Browse Data | Execute SQL

Table: calls

	id	number	date	duration	type	new	name
1	1	7777777777	1388206471836	11	2	0	Tom
2	2	8887775566	1388206593826	5	2	0	
3	3	4444444444	1388211842729	134	2	0	Robert
4	4	6666666666	1388211997835	4	2	0	Amy
5	5	9999999999	1388212023730	1	2	1	James

Extracting SMS/MMS

During the course of investigation, a forensic examiner may be asked to retrieve the text messages that are sent by and delivered to a particular mobile device. Hence, it is important to understand where the details are stored and how to access the data. The `mmssms.db` file which is present under the `/data/data/com.android.providers.telephony/databases` location contains the necessary details. As with call logs, the examiner must ensure that applications capable of messaging are examined for relevant message logs, as follows:

```
C:\android-sdk-windows\platform-tools>adb.exe pull /data/data/com.
android.providers.telephony C:\temp

pull: building file list...
....

-> C:\temp/databases/telephony.db-journal

pull: /data/data/com.android.providers.telephony/databases/mmssms.db ->
C:\temp/databases/mmssms.db

pull: /data/data/com.android.providers.telephony/databases/telephony.db
-> C:\temp/databases/telephony.db

5 files pulled. 0 files skipped.

51 KB/s (160951 bytes in 3.045s)
```

The phone number can be seen under the **address** column and the corresponding text message can be seen under the **body** column, as shown in the following screenshot:

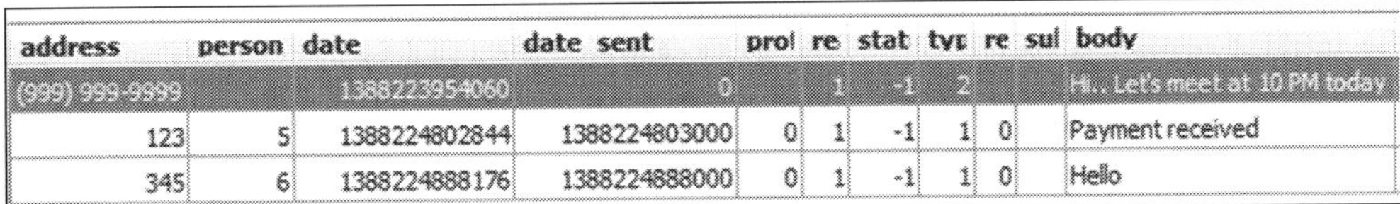

address	person	date	date sent	pro	re	stat	typ	re	sul	body
(999) 999-9999		1388223954060	0		1	-1	2			Hi.. Let's meet at 10 PM today
123	5	1388224802844	1388224803000	0	1	-1	1	0		Payment received
345	6	1388224888176	1388224888000	0	1	-1	1	0		Hello

Calls table in the contacts2.db file

Extracting browser history

Browser history information is one task that is often required to be reconstructed by a forensic examiner. Apart from the default Android Browser, there are different browser applications that can be used on an Android phone, such as Firefox Mobile, Google Chrome, and so on. All of these browsers store their browser history in the SQLite `.db` format. For our example, we are extracting data from the default Android browser to our forensic workstation. This data is located at `/data/data/com.android.browser`. The file named `browser2.db` contains the browser history details. The following screenshot shows the browser data as represented by Oxygen Forensic SQLite Database Viewer. Note that the trial version will hide certain information.

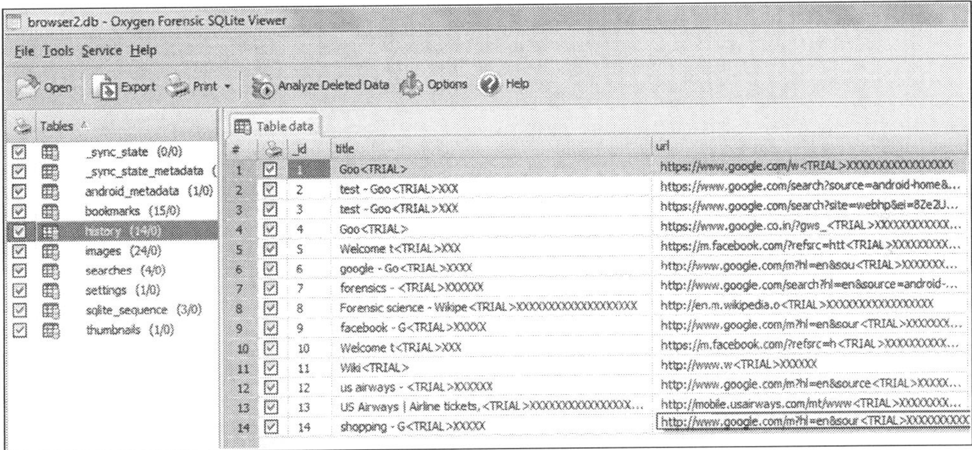

The browser2.db file in Oxygen Forensic SQLite Viewer

Analysis of social networking/IM chats

Social networking and IM chat applications such as Facebook, Twitter, and WhatsApp reveal sensitive data, which could be helpful during the investigation of any case. The analysis is pretty much the same as with any other Android application. Download the data to a forensic workstation and analyze the `.db` files to find out if you can unearth any sensitive information. For example, let's look at the Facebook application and try to see what data can be extracted. First, we extract the `/data/data/com.facebook.katana` folder and navigate to the `databases` folder. The `fb.db` file present under this folder contains information which is associated to the user's account. The `friends_data` table contains information about the friend's names along with their phone numbers, e-mail IDs, and date of birth, as shown in the following screenshot. Similarly, other files can be analyzed to find out if any sensitive information can be gathered.

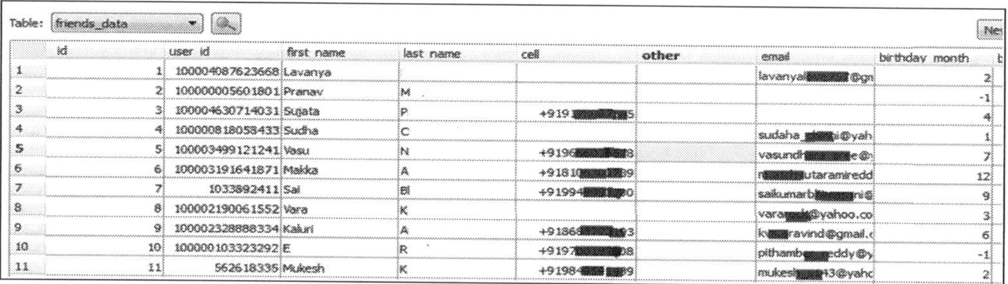

Table: friends_data

	id	user id	first name	last name	cell	other	email	birthday month
1	1	100004087623668	Lavanya				lavanyal[illegible]@gn	2
2	2	100000005601801	Pranav	M				-1
3	3	100004630714031	Sujata	P	+919[illegible]5			4
4	4	100000818058433	Sudha	C			sudaha_[illegible]i@yah	1
5	5	100003499121241	Vasu	N	+9196[illegible]8		vasundh[illegible]e@	7
6	6	100003191641871	Makka	A	+91810[illegible]9		n[illegible]utaramiredd	12
7	7	1033892411	Sai	Bl	+91994[illegible]0		saikumarb[illegible]ni@	9
8	8	100002190061552	Vara	K			vara[illegible]@yahoo.co	3
9	9	100002328888334	Kaluri	A	+91866[illegible]3		k[illegible]ravind@gmail.c	6
10	10	100000103323292	E	R	+9197[illegible]08		pithamb[illegible]eddy@y	-1
11	11	562618335	Mukesh	K	+91984[illegible]9		mukesh[illegible]3@yahc	2

The fb.db file in SQLite browser

Similarly, by analyzing the data present in the `/data/data` folder, information about geo location, calendar events, user notes, and more can be grabbed.

Using content providers

In Android, the data of one application cannot be accessed by another application under normal circumstances. However, Android provides a mechanism through which data can be shared with other applications. This is precisely achieved through the use of content providers. Content providers present data to external applications in the form of one or more tables. These tables are no different from the tables found in a relational database. They can be used by the applications to share data usually through the **URI** addressing scheme. They are used by other applications that access the provider using a provider-client object. During the installation of an app, the user determines whether or not the app can gain access to the requested data (content providers). For instance, contacts, SMS/MMS, calendar, and so on, are examples of content providers.

Hence, by taking advantage of this, we can create an app that can grab all the information from all the available content providers. This is precisely how most of the commercial forensic tools work. The advantage of this method is it can be used on both rooted and non-rooted devices. For our example, we are using **AFLogical**, which takes advantage of the content-provider mechanism to gain access to the information. This tool extracts the data and saves it to an SD card in CSV format. The following steps extract the information from an Android device using AFLogical Open Source Edition 1.5.2:

1. Download AFLogical OSE 1.5.2 from `https://github.com/viaforensics/android-forensics/downloads`.

The AFLogical LE edition is capable of extracting a larger set of information and requires registration with viaForensics using an active law enforcement or government agency e-mail. AFLogical OSE can pull all available MMSes, SMSes, contacts, and call logs.

2. Ensure USB debugging mode is enabled and connect the device to the workstation.
3. Verify that the device is identified by issuing the following command:

```
C:\android-sdk-windows\platform-tools>adb.exe devices
List of devices attached
4df16ac3115d6p18        device
```

4. Save the AFLogical OSE app in the home directory and issue the following command to install it on the device:

```
C:\android-sdk-windows\platform-tools>adb.exe
install AFLogical-OSE_1.5.2.apk
1479 KB/s (28794 bytes in 0.019s)
        pkg: /data/local/tmp/AFLogical-OSE_1.5.2.apk
Success
```

5. Once the application is installed, you can run it directly from the device and click on the **Capture** button present at the bottom of the app, as shown in the following screenshot:

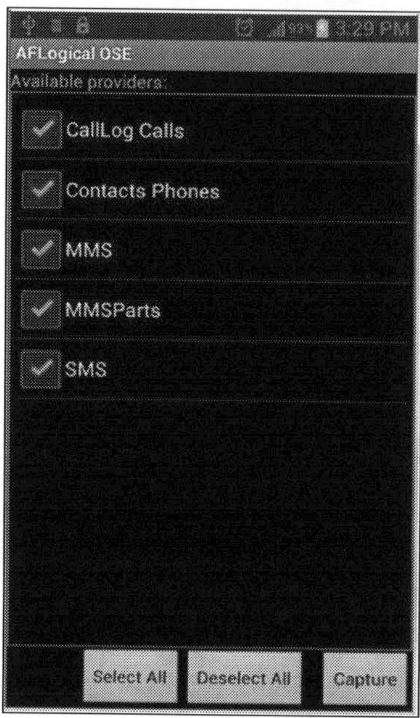

The AFLogical OSE app

6. The app starts extracting data from the respective content providers and once the process is complete, a message will be displayed, as shown in the following screenshot:

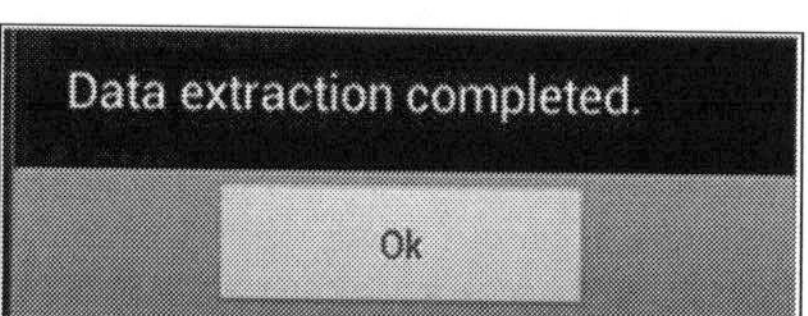

Message displayed after the extraction is complete

7. The extracted data is saved to the SD card of the device in a directory named `forensics`. The extracted information is stored in CSV files, as shown in the following figure. The CSV files can be viewed using any editor.

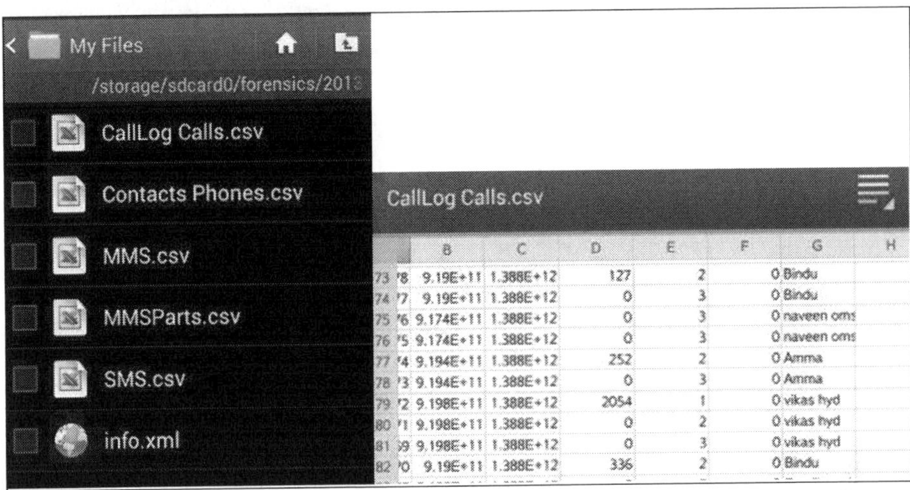

Files extracted using AFLogical OSE

8. The `info.xml` file present in the same directory provides information about the device including the IMEI number, IMSI number, Android version, information about installed applications, and so on.

Other tools that can help during investigation to logically extract data will be covered in *Chapter 11, Android App Analysis and Overview of Forensic Tools*.

Physical data extraction

Android data extraction through physical techniques (hardware-based) mainly involves two methods: JTAG and chip-off. These techniques are usually hard to implement and require great precision and experience to try them on real devices during the course of an investigation. The following sections provide an overview of these techniques.

JTAG

JTAG (Joint Test Action Group) involves using advanced data acquisition methods, which involve connecting to specific ports on the device and instructing the processor to transfer the data stored on the device. By using this method, a full physical image of a device can be acquired. It is always recommended to first try out the logical techniques mentioned earlier as they are easy to implement and require less effort. Examiners must have proper training and experience prior to attempting JTAG as the device may be damaged if handled improperly.

The JTAG process usually involves the following forensic steps:

1. In JTAG, the device **Test Access Ports (TAPs)** are used to access the CPU of the device. Identifying the TAPs is the primary and most important step. TAPs are identified and the connection is traced to the CPU to find out which pad is responsible for each function. Although device manufacturers document resources about the JTAG schematics of a particular device, they are not released for general viewing. A good site for JTAG on an Android device is `http://www.forensicswiki.org/wiki/JTAG_Forensics`.
2. Wire leads are then soldered to appropriate connecter pins and the other end is connected to the device that can control the CPU, as shown in the following image (published by `www.binaryintel.com`). JTAG jigs can be used to forgo soldering for certain devices. The use of a jig or JTAG adapter negates the need to solder, as it connects the TAPs to the CPU.

The JTAG setup

3. Once the preceding steps are complete, power must be applied to boot the CPU. The voltage that must be applied depends on the specifications released by the hardware manufacturer. Do not apply a voltage beyond the number mentioned in the specification.
4. After applying the power, a full binary memory dump of the NAND flash can be extracted.
5. Analyze the extracted data using the forensic techniques and tools learned in this book. A raw `.bin` file will be obtained during the acquisition and most forensic tools support ingestion and analysis of this image format.

It is also important to understand that the JTAG technique should not result in loss of functionality of the device. If reassembled properly, the device should function without any problems. Although the JTAG technique is effective in extracting the data, only experienced and qualified personnel should attempt it. Any error in soldering the JTAG pads or applying a different voltage could damage the device entirely.

Chip-off

Chip-off, as the name suggests, is a technique where the NAND flash chip(s) are removed from the device and examined to extract the information. Hence, this technique will work even when the device is passcode-protected and USB debugging is not enabled. Unlike the JTAG technique where the device functions normally after examination, the chip-off technique usually results in destruction of the device, that is, it is more difficult to reattach the NAND flash to the device after examination. The process of reattaching the NAND flash to the device is called **re-balling** and requires training and practice.

Chip-off techniques usually involve the following forensic steps:

1. All of the chips on the device must be researched to determine which chip contains user data. Once determined, the NAND flash is physically removed from the device. This can be done by applying heat to desolder the chip as shown in the following image (published by `www.binaryintel.com`). This is a very delicate process and must be done with great care as it may result in damaging the NAND flash.

The chip-off technique

2. The chip is then cleaned and repaired to make sure that the connectors are present and functioning.
3. Using specialized hardware device adapters, the chip can now be read. This is done by inserting the chip into the hardware device, which supports the specific NAND flash chip. In this process, raw data is acquired from the chip resulting in a `.bin` file.
4. The data acquired can now be analyzed using forensic techniques and the tools described earlier.

The chip-off technique is most helpful when the device is damaged severely, locked, or otherwise inaccessible. However, the application of this technique requires not only expertise but also costly equipment and tools. There is always a risk of damaging the NAND flash while removing it and hence it is recommended to try out the logical techniques first to extract any data.

Imaging a memory (SD) card

There are many tools available that can image a memory card. The following example uses **WinHex** to create a raw disk image of the SD card. The following is a step-by-step process to image a memory card using the WinHex software.

- **Connecting the memory card**: Remove the SD card from the memory slot and use a card reader to connect the memory card to the forensic workstation.
- **Write protect the card**: Open the disk using WinHex. Navigate to **Options** | **Edit Mode** and select the **write-protected** mode, as shown in the following screenshot. This is to make sure that the device is write protected and no data can be written on it.

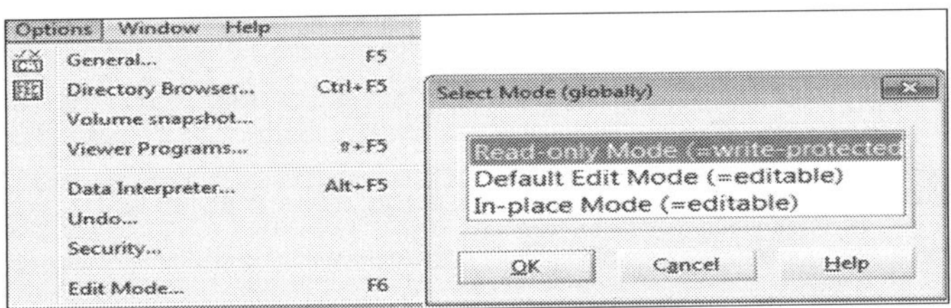

WinHex view of Edit Mode (left) and WinHex Read-only Mode enabled (right)

- **Calculating the hash value**: Calculate the hash value of the memory card to make sure that no changes are made at any point during the investigation. Navigate to **Tools** | **Compute hash** and choose any hashing algorithm.
- **Creating the image of the disk**: Navigate to **File** | **Create Disk Image**, as shown in the following screenshot. Select the **Raw** image option (`.dd`) to create an image. This completes the imaging of the memory card.

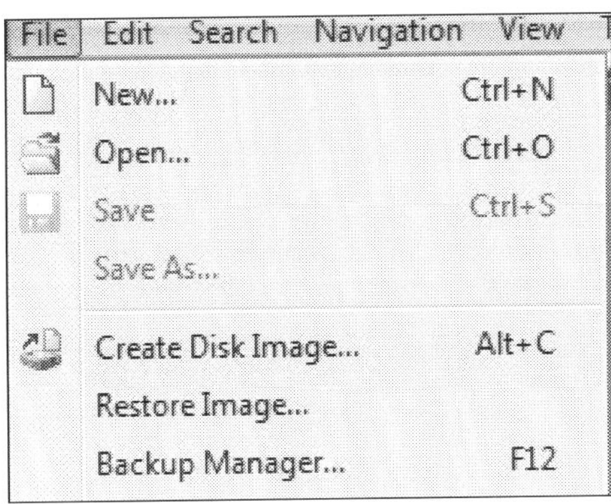

The WinHex disk image option

Summary

Imaging a device is one of the primary steps to ensure that the data on the device is not modified. Once the device is accessible, an examiner can extract the data using manual, logical, or physical data extraction techniques. Logical techniques extract the data by accessing the file system. While the physical techniques access a larger set of data, they are complex and require great expertise to perform. Manual extraction should be performed to validate data or only when one tool is used to create the image. Once the data is acquired, examination and manual extraction follows, as described in the next chapter.

10
Android Data Recovery Techniques

While the data extraction and analysis techniques provide information about various details such as call logs, text messages, and other cellular functions, not all techniques can provide information about the deleted data. It is rare to find a smartphone today that doesn't contain data the user intended to delete. The probability that the deleted data contains sensitive information (which is why the data is deleted in the first place) is high. Hence, data recovery is a crucial aspect of mobile forensics as it helps to unearth the deleted items. This chapter aims to cover various techniques, which can be used by a forensic analyst to recover the data from an Android device.

Data recovery

Data recovery is one of the most significant and powerful aspects of forensic analysis. The ability to recover deleted data can be crucial to crack many civil and criminal cases. From a normal user's point of view, recovering data that has been deleted would usually refer to the operating system's built-in solutions such as the **Recycle Bin** in Windows. While it's true that data can be recovered from these locations, due to an increase in user awareness, these options don't often work. For instance, on a desktop computer, people now use *Shift* + *Delete* as a way to delete a file completely from their desktop.

Data recovery is the process of retrieving deleted data from a device when it cannot be accessed normally. Consider the scenario where a mobile phone has been seized from a terrorist. Wouldn't it be of greatest importance to know which items were deleted by the terrorist? Access to any deleted SMS messages, pictures, dialed numbers, application data, and more can be of critical importance as they often reveal sensitive information. With Android, it is possible to recover most of the deleted data if the device files are properly acquired. However, if proper care is not taken while handling the device, the deleted data could be lost forever. To ensure that the deleted data is not overwritten, it is recommended to keep the following points in mind:

- Do not use the phone for any activity after seizing it. The deleted data exists on the device until the space is needed by some other incoming data. Hence, the phone must not be used for any sort of activity so as to prevent the data from being overwritten.
- Even when the phone is not used, without any intervention from our end, data can be overwritten. For instance, an incoming SMS would automatically occupy the space, which could overwrite the data marked for deletion. To prevent occurrence of such events, the examiner should follow the forensic handling methods described in the previous chapters. The easiest solution is to place the device in airplane mode, disable all connectivity options on the device, or turn the device off. This prevents the delivery of any new messages.

Recovering the deleted files

All Android file systems have metadata containing information about the hierarchy of files, filenames, and so on. Deletion will not really erase the data but remove the file system metadata. When text messages or any other files are deleted from the device, they are just made invisible to the user but the files are still present on the device. Essentially, the files are simply marked for deletion, but reside on the file system until being overwritten. Recovering deleted data from an Android device involves two scenarios: recovering data that is deleted from the SD card, such as pictures, videos, application data, and more, and recovering data that is deleted from the internal memory of the device. The following sections cover the techniques that can be used to recover deleted data from both the SD card and internal memory of the Android device.

Recovering deleted data from an SD card

Data present on SD cards can reveal a lot of information for forensic investigators. SD cards are capable of storing pictures and videos taken by the phone's camera, voice recordings, application data, cached files, and more. Essentially, anything that can be stored on a computer hard drive can be stored on an SD card as much as the available space allows. Recovering the deleted data from an external SD card is a straightforward process. SD cards can be mounted as an external mass storage device and forensically acquired using standard digital forensic methods as discussed in *Chapter 9, Android Data Extraction Techniques*. The device should never be mounted on a computer to copy the files as the unallocated space will be missed. As mentioned in the previous chapters, SD cards in Android devices often use the FAT32 file system. The main reason for this is that the FAT32 file system is widely supported in most operating systems including Windows, Linux, and Mac OS X. The maximum file size on a FAT32 formatted drive is around 4 GB. With increasingly high resolution formats now available, this limit is commonly reached. Apart from this, FAT32 can be used on partitions that are less than 32 GB in size. Hence, the exFAT file system, which overcomes these problems, is now being used in some of the devices.

To recover the deleted files from an SD card, you can use any of the available forensic tools such as the **Remo Recover for Android** tool. The following is a step-by-step process to recover the deleted files from an SD card using Remo Recovery for Android:

1. Download the software from `http://www.remosoftware.com/remo-recover-for-android`. Next, install the software and launch it. From the main screen, select the appropriate file recovery mode. The tool tries to recognize the Android device and displays the following screen, once the device is successfully detected. Note, the Android device must be able to connect via USB debugging or the device may not be detected.

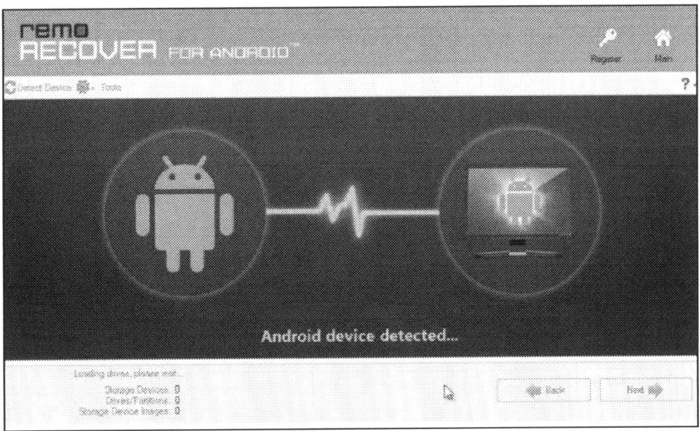

Android recovery—device detection

2. The tool presents you with a list of storage devices available, as shown in the following screenshot. Select the storage device from the list and proceed.

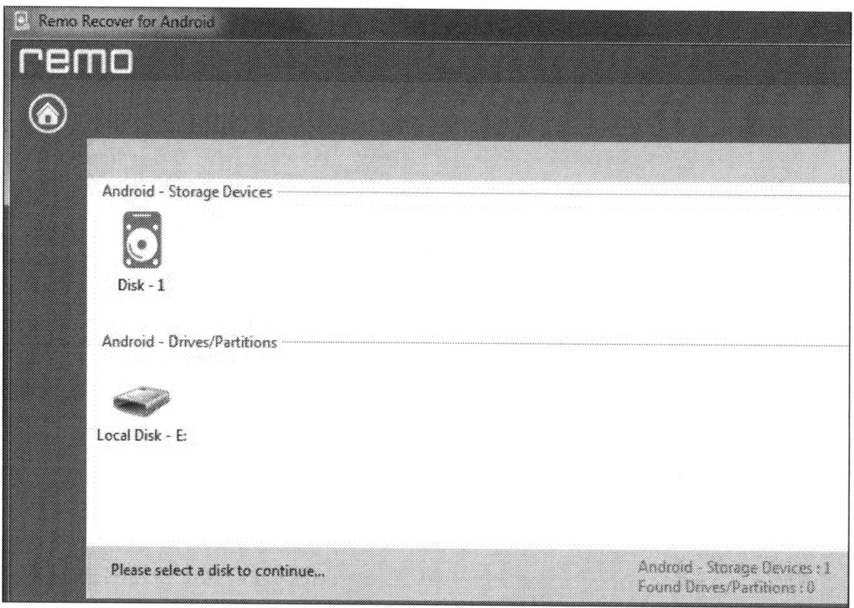

The list of storage devices available

3. Select the type of file to be recovered or select all and proceed further.
4. Once the recovery process is complete, a list of the extracted files will be provided as shown in the following screenshot:

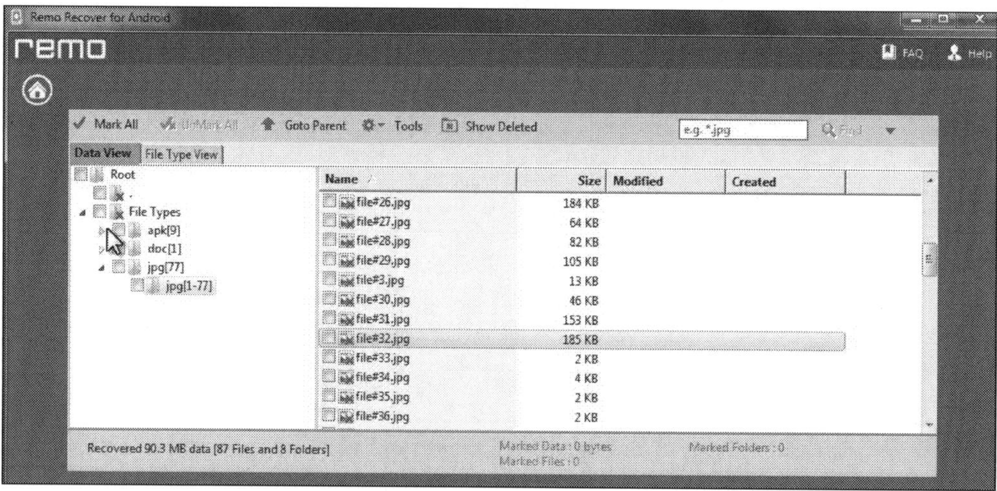

Recovered files list

Examiners must understand that Android devices might use space on the SD card to cache application data, therefore it is important to make sure that as much data as possible is obtained from the device prior to removing the SD card. It is recommended to acquire the SD card through the device as well as separately to ensure all data is obtained. To achieve the SD card image, `dd` through adb can be used while the device is running to obtain an image of the SD card of the device if the device cannot be powered off due to possible evidence running in the memory. A memory capture can be obtained on the Android device should data actively be running in the memory be relevant to the investigation. Tools such as **LiME** can be used to complete the memory capture. LiME can be accessed on the following site: `https://code.google.com/p/lime-forensics/`.

It is also recommended to check if the device has any backup applications or files installed. The initial release of Android did not include a mechanism for the users to back up their personal data. Hence, several backup applications were used extensively by the users. By using the apps, users have the ability to back up their data either to the SD card or to the cloud. For example, the **Super Backup** app contains the options to back up call logs, contacts, SMS, and more as shown in the following screenshot:

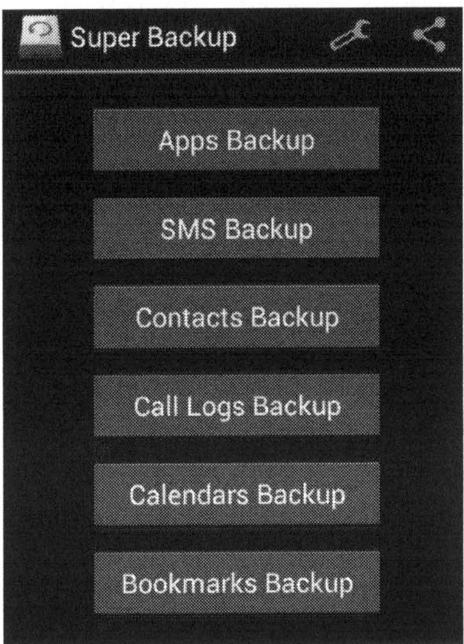

The Super Backup Android app

Upon detection of a backup application, the forensic examiners must attempt to determine where the data is stored. The data saved in a backup may contain important information and thus looking for any third-party backup app on the device would be very helpful.

Recovering data deleted from internal memory

Recovering files which are deleted from Android's internal memory (such as SMS, contacts, app data, and more) is not supported by all analytical tools and may require manual carving. Unlike some media containing common file systems such as SD cards, the file system may not be recognized and mounted by forensic tools. Also, the examiner cannot get access to the raw partitions of the internal memory of an Android phone unless the phone is rooted. The following are some of the other issues the examiner may face when attempting to recover data from the internal memory on Android devices:

- To get access to the internal memory you can try to root the phone. However, the rooting process might involve writing some data to the `/data` partition and this process could overwrite the data of value on the device.
- Unlike SD cards, the internal file system here is not FAT32 (which is widely supported by forensic tools). The internal file system could be YAFFS2 (in older devices), EXT3, EXT4, RFS, or something proprietary built to run on Android. Therefore, many of the recovery tools designed for use with Windows file systems won't work.
- Application data on Android devices is commonly stored in the SQLite format. While most forensic tools provide access to the database files, they may have to be exported and viewed in a native browser. The examiner must examine the raw data to ensure that the deleted data is not overlooked by the forensic tool.

The discussed reasons make it difficult, but not impossible, to recover the deleted data from the internal memory. The internal memory of Android devices holds a bulk of the user data and the possible keys to your investigation. As previously mentioned, the device must be rooted in order to access the raw partitions. Most of the Android recovery tools on the market do not highlight the fact that they work only on rooted phones. Let us now see how we can recover deleted data from an Android phone.

Recovering deleted files by parsing SQLite files

Android uses SQLite files to store most data. Data related to text messages, e-mails, and certain app data is stored in SQLite files. SQLite databases can store deleted data within the database itself. Files marked for deletion by the user no longer appear in the active SQLite database files. Therefore, it is possible to recover the deleted data such as text messages, contacts, and more. There are two areas within a SQLite page that can contain deleted data: unallocated blocks and free blocks. Most of the commercial tools that recover deleted data scan the unallocated blocks and free blocks of the SQLite pages. Parsing the deleted data can be done using the available forensic tools such as **Oxygen Forensics SQLite Viewer**. The trial version of the SQLite Viewer can be used for this purpose; however, there are certain limitations on the amount of data that you can recover. You can write your own script to parse the files for deleted content and for that you need to have a good understanding about the SQLite file format. The link `http://www.sqlite.org/fileformat.html` is a good place to start with. If you do not want to reinvent and want to reuse the existing scripts, you can try the available open source Python scripts (`http://az4n6.blogspot.in/2013/11/python-parser-to-recover-deleted-sqlite.html`) to parse the SQLite files for deleted records.

For our example, we will recover deleted SMSes from an Android device. Recovering deleted SMSes from an Android phone is quite often requested as part of forensic analysis on a device mainly because text messages contain data, which can reveal a lot of information. There are different ways to recover deleted text messages on an Android device. First, we need to understand where the messages are being stored on the device. In *Chapter 9, Android Data Extraction Techniques*, we explained the important locations on the Android device where user data is stored. Here is a quick recap of this:

- Every application stores its data under the `/data/data` folder (again, this requires root access to acquire data)
- The files under the location `/data/data/com.android.providers.telephony/databases` contain details about SMS/MMS

Under the preceding mentioned location, text messages are stored in a SQLite database file, which is named `mmssms.db`. Deleted text messages can be recovered by examining this file. Here are the steps to recover deleted SMSes using the `mmssms.db` file:

1. On the Android device, enable the USB debugging mode and connect the device to the forensic workstation. Using the adb command-line tool, extract the `databases` folder present under the location `/data/data/` by issuing the adb pull command:

   ```
   C:\android-sdk-windows\platform-tools>adb.exe pull
   /data/data/com.android.providers.telephony/databases C:\temp
   ```

```
pull: building file list...
pull:
/data/data/com.android.providers.telephony/databases/mmssms.db
-journal -> C:\temp/mmssms.db-journal
pull:
/data/data/com.android.providers.telephony/databases/telephony
.db-journal -> C:\temp/telephony.db-journal
pull:
/data/data/com.android.providers.telephony/databases/mmssms.db
-> C:\temp/mmssms.db
pull:
/data/data/com.android.providers.telephony/databases/telephony
.db -> C:\temp/telephony.db
4 files pulled. 0 files skipped.
53 KB/s (160848 bytes in 2.958s)
```

Once the files are extracted to the local machine, use the Oxygen Forensics SQLite Viewer tool to open the `mmssms.db` file.

2. Click on the table named `sms` and observe the current message under the **Tables** data tab in the tool.
3. One way to view the deleted data is by clicking on the **Blocks containing deleted data** tab, as shown in the following screenshot:

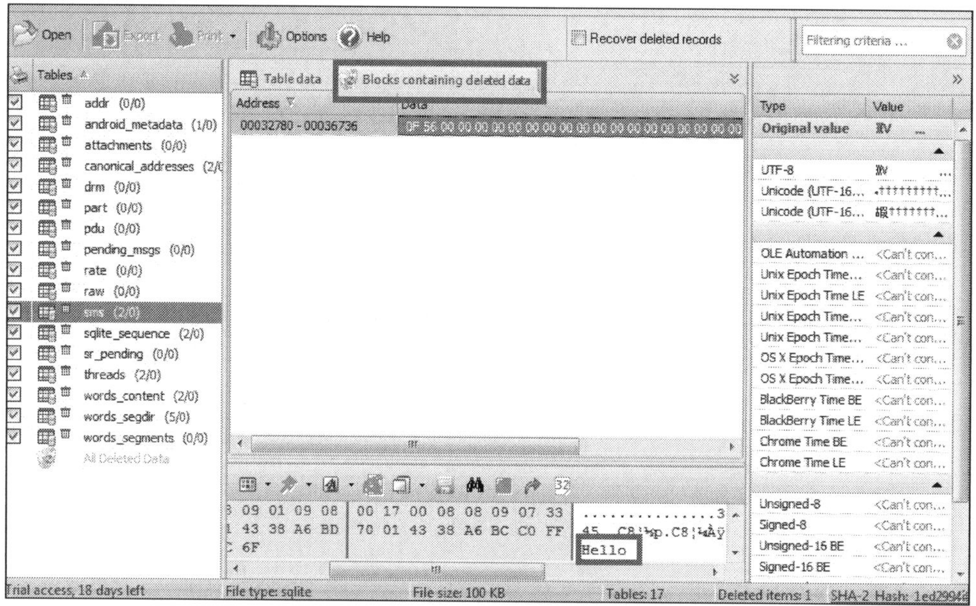

Recovering deleted SMS messages

Similarly, other data residing on Android devices which store data in SQLite files can be recovered by parsing for deleted content. When the preceding method doesn't provide access to the deleted data, the examiner should look at the file in raw hex file for data marked as deleted, which can be manually carved and reported.

Recovering files using file-carving techniques

File carving is an extremely useful method in forensics because it allows for data that has been deleted or hidden to be recovered for analysis. In simple terms, file carving is the process of reassembling computer files from fragments in the absence of file system metadata. In file carving, specified file types are searched for and extracted across the binary data to create a forensic image of a partition or an entire disk. File carving recovers files from the unallocated space in a drive based merely on file structure and content without any matching file system metadata. Unallocated space refers to the part of the drive that no longer holds any file information as pointed by the file system structures such as the file table.

Files can be recovered or reconstructed by scanning the raw bytes of the disk and reassembling them. This can be done by examining the header (the first few bytes) and footer (the last few bytes) of a file.

File-carving methods are categorized based on the underlying technique in use. The header-footer carving method relies on recovering the files based on the header and footer information. For instance, the JPEG files start with `0xffd8` and end with `0xffd9`. The locations of the header and footer are identified and everything between those two endpoints is carved. Similarly, the carving method based on the file structure uses the internal layout of a file to reconstruct the file. But the traditional file-carving techniques such as the ones we've already explained may not work if the data is fragmented. To overcome this, new techniques such as smart carving use the fragmentation characteristics of several popular file systems to recover the data.

Once the phone is imaged, it can be analyzed using tools such as **Scalpel**. Scalpel is a powerful open source utility to carve files. This tool analyzes the block database storage and identifies the deleted files and recovers them. Scalpel is file system independent and is known to work on various file systems including FAT, NTFS, EXT2, EXT3, HFS, and more. The following steps explain how to use Scalpel on an Ubuntu workstation:

1. Install Scalpel on the Ubuntu workstation using the command `sudo apt-get install scalpel`.

2. The `scalpel.conf` file present under the `/etc/scalpel` directory contains information about the supported file types, as shown in the following screenshot:

```
# GRAPHICS FILES
#---------------------------------------------------------------------
#
#
# AOL ART files
#       art     y       150000  \x4a\x47\x04\x0e        \xcf\xc7\xcb
#       art     y       150000  \x4a\x47\x03\x0e        \xd0\xcb\x00\x00
#
# GIF and JPG files (very common)
#       gif     y       5000000         \x47\x49\x46\x38\x37\x61        \x00\x3b
#       gif     y       5000000         \x47\x49\x46\x38\x39\x61        \x00\x3b
#       jpg     y       200000000       \xff\xd8\xff\xe0\x00\x10        \xff\xd9
#
#
# PNG
#       png     y       20000000        \x50\x4e\x47?   \xff\xfc\xfd\xfe
#
#
# BMP   (used by MSWindows, use only if you have reason to think there are
#       BMP files worth digging for. This often kicks back a lot of false
#       positives
```

The scalpel configuration file

This file needs to be modified in order to mention the files that are related to Android. A sample `scalpel.conf` file can be downloaded from the link `https://viaforensics.com/resources/tools/#android`. You can also uncomment the files and save the `conf` file to select file types of your choice. Once this is done, replace the original `conf` file with the one that is downloaded.

3. Scalpel needs to be run along with the preceding configuration file on the `dd` image being examined. You can run the tool using the command shown in the following screenshot, by inputting the configuration file and the `dd` file. Once the command is run, the tool starts to carve the files and build them accordingly.

```
File Edit View Search Terminal Help
unigeek@ubuntu:~$ scalpel -c /home/unigeek/Desktop/scalpel-android.conf /home/un
igeek/Desktop/userdata.dd -o /home/unigeek/Desktop/rohit
Scalpel version 1.60
Written by Golden G. Richard III, based on Foremost 0.69.

Opening target "/home/unigeek/Desktop/userdata.dd"

Image file pass 1/2.
/home/unigeek/Desktop/userdata.dd: 100.0% |*************|     3.9 MB    00:00 ETA
Allocating work queues...
Work queues allocation complete. Building carve lists...
Carve lists built.  Workload:
gif with header "\x47\x49\x46\x38\x37\x61" and footer "\x00\x3b" --> 0 files
gif with header "\x47\x49\x46\x38\x39\x61" and footer "\x00\x3b" --> 2 files
jpg with header "\xff\xd8\xff\xe0\x00\x10" and footer "\xff\xd9" --> 71 files
jpg with header "\xff\xd8\xff\xe1" and footer "\x7f\xff\xd9" --> 1 files
png with header "\x50\x4e\x47\x3f" and footer "\xff\xfc\xfd\xfe" --> 0 files
png with header "\x89\x50\x4e\x47" and footer "" --> 71 files
sqlitedb with header "\x53\x51\x4c\x69\x74\x65\x20\x66\x6f\x72\x6d\x61\x74" and
footer "" --> 0 files
email with header "\x46\x72\x6f\x6d\x3a" and footer "" --> 0 files
doc with header "\xd0\xcf\x11\xe0\xa1\xb1\x1a\xe1\x00\x00" and footer "\xd0\xcf\
x11\xe0\xa1\xb1\x1a\xe1\x00\x00" --> 0 files
doc with header "\xd0\xcf\x11\xe0\xa1\xb1" and footer "" --> 0 files
htm with header "\x3c\x68\x74\x6d\x6c" and footer "\x3c\x2f\x68\x74\x6d\x6c\x3e"
 --> 1 files
pdf with header "\x25\x50\x44\x46" and footer "\x25\x45\x4f\x46\x0d" --> 0 files
pdf with header "\x25\x50\x44\x46" and footer "\x25\x45\x4f\x46\x0a" --> 0 files
wav with header "\x52\x49\x46\x46\x3f\x3f\x3f\x3f\x57\x41\x56\x45" and footer ""
 --> 0 files
amr with header "\x23\x21\x41\x4d\x52" and footer "" --> 0 files
```

Running the Scalpel tool on a dd file

4. The output folder specified in the preceding command now contains lists of folders based on the file types, as shown in the following screenshot. Each of these folders contains data based on the folder name. For instance, `jpg 2-0` contains files related to the `.jpg` extension that has been recovered.

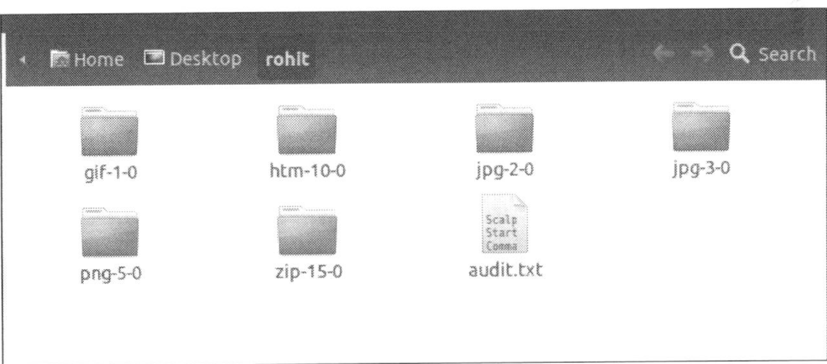

Output folder after running the Scalpel tool

5. As shown in the preceding screenshot, each folder contains recovered data from the Android device, such as images, PDF files, ZIP files, and more. While some pictures are recovered completely, some are not recovered to a full extent, as shown in the following screenshot:

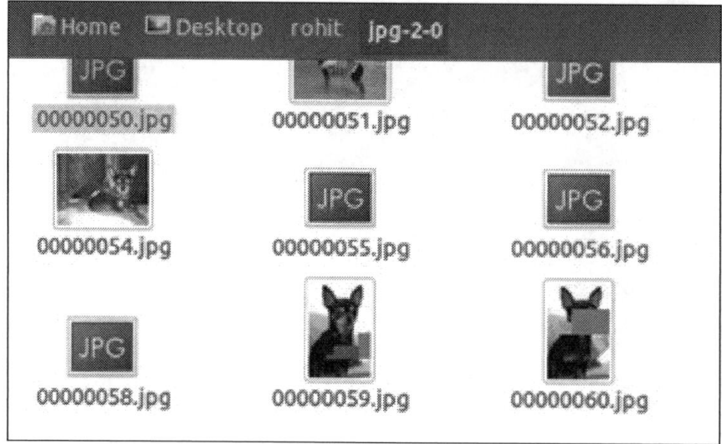

Recovered data using the Scalpel tool

Applications such as **DiskDigger** can be installed on Android devices to recover different types of files from both the internal memory and SD cards. Applications such as DiskDigger include support for JPG files, MP3 and WAV audio, MP4 and 3GP video, raw camera formats, Microsoft Office files (DOC, XLS, and PPT), and more. However, as mentioned earlier, the application requires root privileges on the Android device in order to recover the content from the internal memory. Thus, file-carving techniques play a very important role in recovering important deleted files from the device's internal memory.

You can also restore the contacts on the device using the **Restore Contacts** option through the Google account configured on the device. This would work if the user of the device has previously synced their contacts using the **Sync Settings** option available in Android. This option synchronizes the contacts and other details and would store them in the cloud. A forensic examiner with legal authority or proper consent can restore the deleted contacts if they can get access to the Google account configured on the device. Once the account is accessed, perform the following steps to restore the data:

1. Log in to the Gmail account.
2. Click on **Gmail** in the top-left corner and select **Contacts**, as shown in the following screenshot:

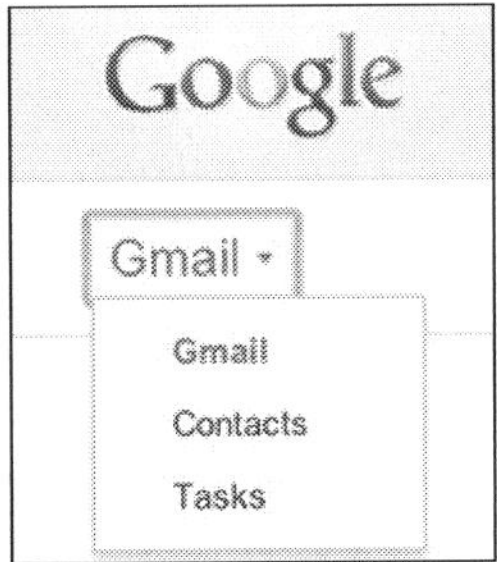

The Contacts menu in Gmail

3. Click on **More**, which is present above the contacts list.
4. Click on **Restore Contacts** and the following screen appears:

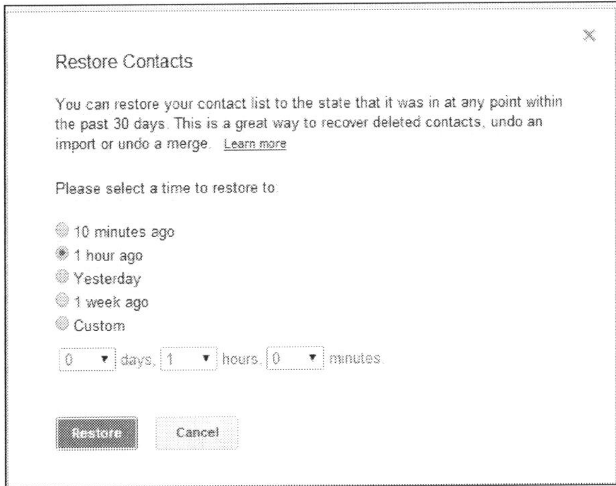

The Restore Contacts dialog box

5. Now, you can restore the contact list to the state that it was in at any point within the past 30 days using this technique.

Summary

Recovery of the deleted data on Android devices depends on various factors which heavily rely on access to the data residing in the internal memory and SD card. While the recovery of deleted items from external storage such as an SD card is easy, recovery of deleted items from the internal memory takes considerable effort. SQLite file parsing and file-carving techniques are two methods to recover deleted data extracted from an Android device. The next chapter discusses Android forensic tools that can be helpful in extracting and acquiring data from Android devices. Both open source and commercial methods will be discussed.

11 Android App Analysis and Overview of Forensic Tools

Third-party applications are commonly used by smartphone users. Android users download and install several apps from app stores such as Android Market and Google Play. During forensic investigations, it is often helpful to perform an analysis of these apps to retrieve valuable data and to detect any malware. For instance, a photo vault app might lock sensitive images present on the device. Hence, it would be of great significance to have the knowledge to identify the passcode for the photo vault app. While the data extraction and data recovery techniques discussed in earlier chapters provide access to valuable data, app analysis would help us to gain information about the specifics of an application, such as preferences and permissions. This chapter covers the techniques to reverse engineer an Android application and also throws light on some available forensic tools that may be extremely helpful during forensic acquisition and analysis.

Android app analysis

On Android, everything the user interacts with is an application. While some apps are preinstalled by the device manufacturer, some apps are downloaded and installed by the user. Depending on the type of application, most of these apps store sensitive information on the internal memory or the SD card on the device. Using the forensic techniques described earlier, it is possible to get access to the data stored by these applications. However, a forensic examiner needs to develop the necessary skills to convert the available data into useful data. This is achieved when you have a comprehensive understanding of how the application handles data.

The examiner may need to deal with applications that stand as a barrier to accessing required information. For instance, take the case of the gallery on a phone locked by an app locker application. In this case, in order to access the pictures and videos stored on the gallery, you first need to enter the passcode to the app locker. Hence, it would be interesting to know how the app locker app stores the password on the device. You might look into the sqlite database files, but if they are encrypted, then it's hard to even predict that it's a password. Reverse engineering applications would be helpful in such cases where you want to better understand the application and how the application stores the data.

Reverse engineering Android apps

To state it in simple terms, reverse engineering is the process of retrieving source code from an executable. Reverse engineering an Android app is done in order to understand the functioning of the app, data storage, security mechanisms in place, and more. Before we proceed to learn how to reverse engineer an Android app, here is a quick recap of the Android apps:

- All the applications that are installed on the Android device are written in the Java programming language.
- When a Java program is compiled, we get bytecode. This is sent to a dex compiler, which converts it into a **Dalvik** bytecode.
- Thus, the class files are converted to dex files using dx tool. Android uses something called **Dalvik virtual machine** (**DVM**) to run its applications.
- JVM's bytecode consists of one or more class files depending on the number of Java files that are present in an application. Regardless, a Dalvik bytecode is composed of only one dex file.

Thus, the dex files, XML files, and other resources that are required to run an application, are packaged into an Android package file (an APK file). These APK files are simply a collection of items within a ZIP file. Therefore, if you rename an APK extension file as `.zip`, then you will be able to see the contents of the file. But before that, you need to get access to the APK file of the application that is installed on the phone. Here is how the APK file corresponding to an application can be accessed.

Extracting an APK file from an Android device

Apps that come preinstalled with the phone are stored in the `/system/app` directory. Third-party applications that are downloaded by the user are stored in the `/data/app` folder. The following method helps you to gain access to the APK files on the device and works on both rooted and non-rooted devices:

1. Identify the package name of the app by issuing the following command:

   ```
   C:\android-sdk-windows\platform-tools>adb.exe shell pm list
   packages
   package:android
   package:android.googleSearch.googleSearchWidget
   package:com.android.MtpApplication
   package:com.android.Preconfig
   package:com.android.apps.tag
   package:com.android.backupconfirm
   package:com.android.bluetooth
   package:com.android.browser
   package:com.android.calendar
   package:com.android.certinstaller
   package:com.android.chrome
   ...
   ```

   As shown in the preceding command lines, the list of package names is displayed. Try to find a match between the app in question and the package name. Usually, the package names are very much related to the app names. Alternatively, you can use the Android Market or Google Play to identify the package name easily. The URL for an app in Google Play contains the package name as shown in the following screenshot:

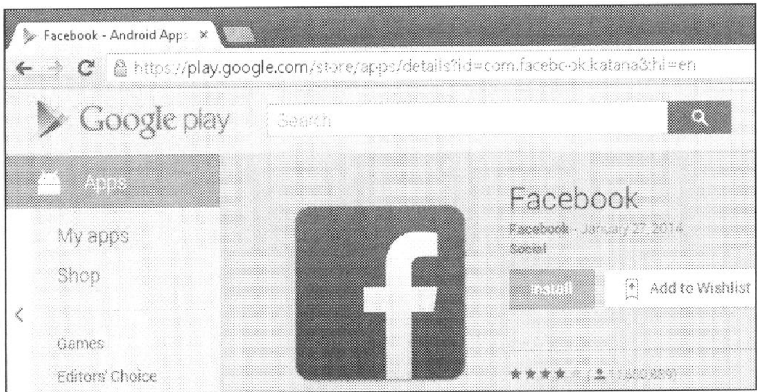

Facebook App in Google Play Store

2. Identify the full path name of the APK file for the desired package by issuing the following command:

```
C:\android-sdk-windows\platform-tools>adb.exe shell pm path com.
android.chrome
package:/data/app/com.android.chrome-2.apk
```

3. Pull the APK file from the Android device to the forensic workstation using the adb pull command:

```
C:\android-sdk-windows\platform-tools>adb.exe pull /data/app/com.
android.chrome-2.apk C:\temp
3493 KB/s (30943306 bytes in 8.649s)
```

You can also use applications such as **ES Explorer** to get the APK file of an Android application. Now let's analyze the contents of an APK file. An Android package is a container for an Android app's resources and executables. It's a zipped file that contains the following files:

- `AndroidManifest.xml`: This contains information about the permissions and more
- `classes.dex`: This is the class file converted to a dex file by the dex compiler
- `Res`: The application's resources, such as the image files, sound files, and more, are present in this directory
- `Lib`: This contains native libraries that the application may use
- `META-INF`: This contains information about the application's signature and signed checksums for all the other files in the package.

Once the APK file is obtained, you can proceed to reverse engineer the Android application.

Steps to reverse engineer Android apps

The APK files can be reverse engineered in different ways to get the original code. The following is one method that uses the **dex2jar** and **JD-GUI** tools to gain access to the application code. For our example, we will examine the `com.twitter.android-1.apk` file. The following are the steps to successfully reverse engineer the APK file:

1. Rename the `apk` extension with `zip` to see the contents of the file. Rename the `com.twitter.android-1.apk` file to `twitter.android-1.zip`, and extract the contents of the file using any file archiver application. The following screenshot shows the files extracted from the original file `twitter.android-1.zip`:

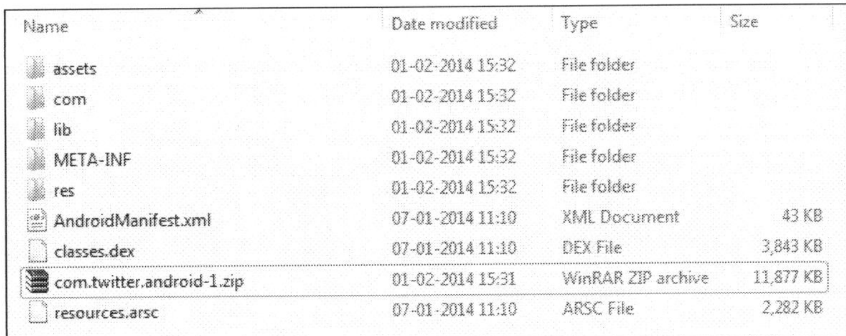

Name	Date modified	Type	Size
assets	01-02-2014 15:32	File folder	
com	01-02-2014 15:32	File folder	
lib	01-02-2014 15:32	File folder	
META-INF	01-02-2014 15:32	File folder	
res	01-02-2014 15:32	File folder	
AndroidManifest.xml	07-01-2014 11:10	XML Document	43 KB
classes.dex	07-01-2014 11:10	DEX File	3,843 KB
com.twitter.android-1.zip	01-02-2014 15:31	WinRAR ZIP archive	11,877 KB
resources.arsc	07-01-2014 11:10	ARSC File	2,282 KB

Extracted files of an APK file

2. The `classes.dex` file discussed in the earlier sections can be accessed after extracting the contents of the APK file. This dex file needs to be converted to a class file of Java. This can be done using the dex2jar tool.
3. Download the dex2jar tool from `https://code.google.com/p/dex2jar/`, and drop the `classes.dex` file into the dex2jar tools directory and issue the following command:

```
C:\Users\Rohit\Desktop\Training\Android\dex2jar-0.0.9.15>d2j-
dex2jar.bat classes.dex

dex2jar classes.dex -> classes-dex2jar.jar
```

4. The preceding command, when successfully run, creates a new file `classes-dex2jar.jar` in the same directory as shown in the following screenshot:

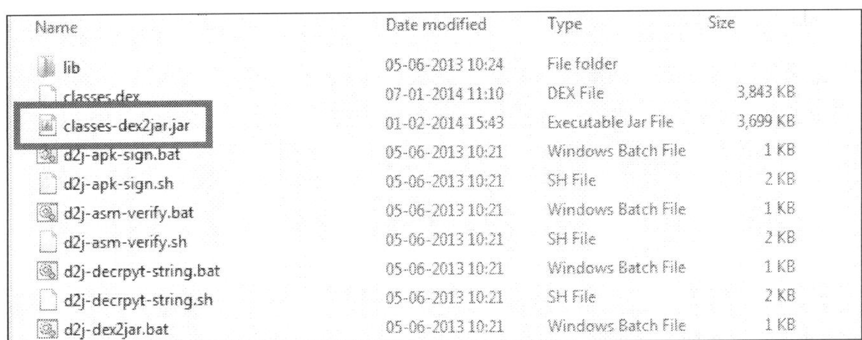

Name	Date modified	Type	Size
lib	05-06-2013 10:24	File folder	
classes.dex	07-01-2014 11:10	DEX File	3,843 KB
classes-dex2jar.jar	01-02-2014 15:43	Executable Jar File	3,699 KB
d2j-apk-sign.bat	05-06-2013 10:21	Windows Batch File	1 KB
d2j-apk-sign.sh	05-06-2013 10:21	SH File	2 KB
d2j-asm-verify.bat	05-06-2013 10:21	Windows Batch File	1 KB
d2j-asm-verify.sh	05-06-2013 10:21	SH File	2 KB
d2j-decrpyt-string.bat	05-06-2013 10:21	Windows Batch File	1 KB
d2j-decrpyt-string.sh	05-06-2013 10:21	SH File	2 KB
d2j-dex2jar.bat	05-06-2013 10:21	Windows Batch File	1 KB

The classes-dex2jar.jar file created by the dex2jar tool

5. To view the contents of this jar file, you can use a tool such as **JD-GUI**. As shown in the following screenshot, the files present in an Android application and the corresponding code can be seen:

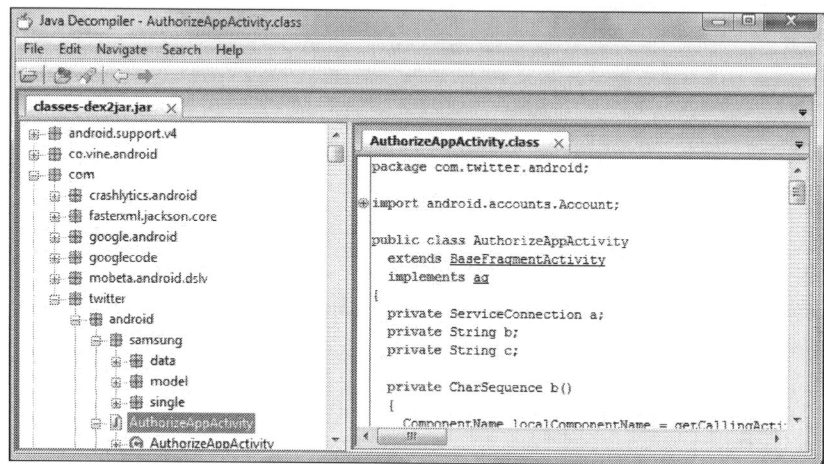

The JD-GUI tool

Once we get access to the code, it is easy to analyze how the application stores the values, permissions, and more information that may be helpful to bypass certain restrictions. When malware is found on a device, this method to decompile and analyze the application may prove useful, as it will show what is being accessed by the malware and clues to where the data is being sent. The method in the preceding screenshot is the best way to determine how malware is affecting the Android device.

Forensic tools overview

It is important for an examiner to understand how a forensic tool acquires and analyzes data to ensure nothing is missed and that the data is being decoded correctly. While manual extraction and analysis is useful, a forensic examiner may need the help of tools to accomplish the tasks involved in mobile device forensics. Forensic tools not only save time, but also make the process a lot easier. The following section describes four important tools that are widely used during forensic acquisition and the analysis of an Android device.

The AFLogical tool

AFLogical is an Android forensics tool developed by **viaForensics**. This tool performs logical acquisition of any Android device running either Android 1.5 or later versions. It allows the extracted data to be saved to the examiner's SD Card in CSV format. There are two editions in this tool: AFLogical **Open Source Edition** (**OSE**) and AFLogical **Law Enforcement** (LE).

AFLogical Open Source Edition

AFLogical Open Source Edition is free open source software. It pulls all available MMS, SMS, contacts, and call logs from the Android device. AFLogical OSE is also built into Santoku-Linux, the open source, community-driven OS dedicated to mobile forensics, mobile malware, and mobile security. The concepts behind AFLogical OSE were mentioned in *Chapter 9, Android Data Extraction Techniques*. This edition can also be used on Santoku-Linux by performing the following steps:

1. Navigate to **Santoku** | **Device Forensics** | **AFLogical OSE**, as shown in the following screenshot:

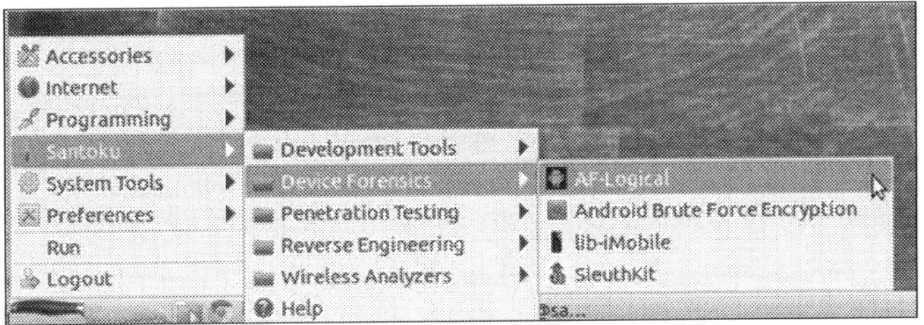

AFLogical in Santoku Linux

2. To install AFLogical OSE onto the device, connect the Android device via USB. If you are using Santoku-Linux in a VM, make sure you connect the Android device to the guest VM.
3. Install the application to your device as follows:

```
aflogical-ose

634 KB/s (28794 bytes in 0.044s)
pkg: /data/local/tmp/AFLogical-OSE_1.5.2.apk
Success

Starting: Intent { cmp=com.viaforensics.android.aflogical_ose/com.
viaforensics.android.ForensicsActivity }

Press enter to pull /sdcard/forensics into ~/aflogical-data/
```

4. On the Android device, select the items you wish to extract and click on **Capture**.
5. Next, press *Enter* in the Linux workstation. This will extract the data from your Android device to the mounted SD card in `~/aflogical-data`.
6. The data is stored in a folder labeled with the date and time of the extraction, as shown in the following screenshot referenced from `https://santoku-linux.com/`:

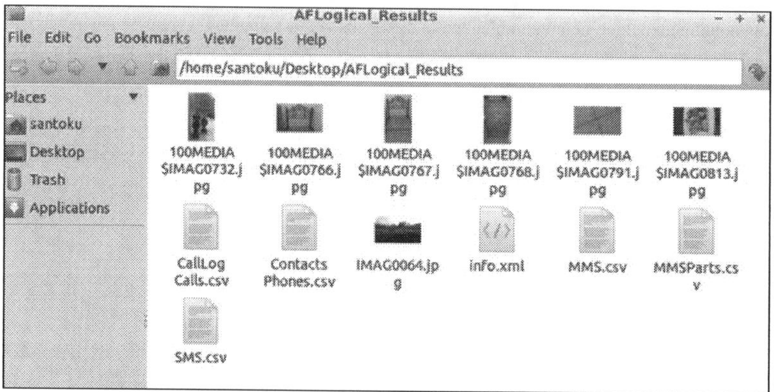

The AFLogical results

7. The extracted data, such as call logs, SMS, contacts, and more, can be . accessed by browsing this folder.

AFLogical Law Enforcement (LE)

According to viaForensics, to download AFLogical LE, you must register with viaForensics using an active law enforcement or government agency e-mail. This edition is able to pull all logical data from an Android device, including the following:

- Browser bookmarks
- Browser searches
- Calendar attendees
- Calendar events
- Calendar extended properties
- Calendar reminders
- Calendars
- CallLog calls

- Contact methods
- Contact extensions
- Contact groups
- Contact organizations
- Contact phones
- Contact settings
- External image media
- External image thumb media
- External media external videos
- IM account
- IM accounts
- IM chats
- IM contacts provider (IM contacts)
- IM invitations
- IM messages
- IM providers
- IM provider settings
- Internal image media
- Internal image thumb media
- Internal Videos and Maps-Friends
- Maps-Friends contacts
- Maps-Friends extra
- MMS
- MmsPartsProvider (MMSParts)
- Notes
- People
- Phone storage deleted by people (HTC Incredible)
- Search history
- SMS
- Social contracts activities

Cellebrite – UFED

Currently, Cellebrite UFED offers several products that support data acquisition and analysis of Android devices. Cellebrite is a popular commercial tool that provides the examiner with both logical and physical acquisition support as well as an analytical platform to examine data. **Cellebrite Physical Analyzer**, the analytical platform, allows the examiner to keyword search, bookmark, carve data, and create customized reports to support their investigation.

Physical extraction

The following steps need to be followed to extract information from a Samsung Android device using UFED Touch. Before the extraction process starts, make sure that the phone is fully charged.

1. In the UFED Touch menu, select **Physical Extraction**, as shown in the following screenshot:

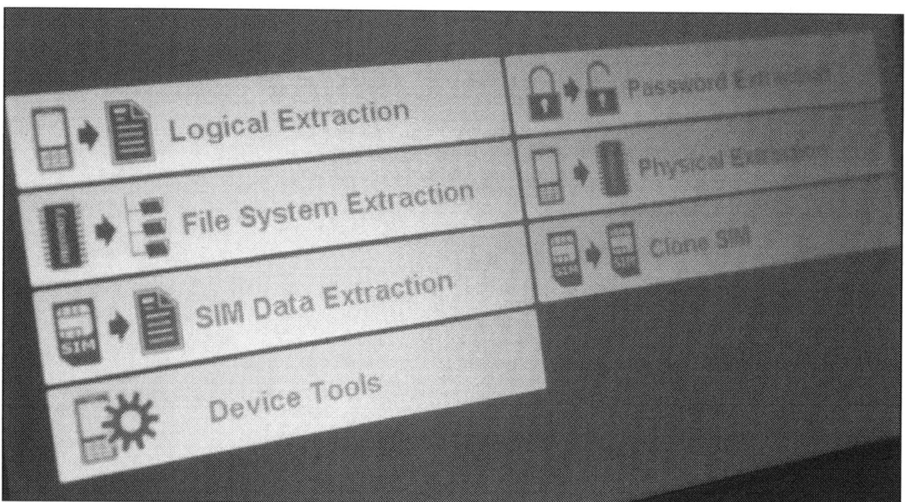

The UFED Touch main menu

2. In the vendor list, select the name of the device manufacturer as shown in the following screenshot (for example, Samsung):

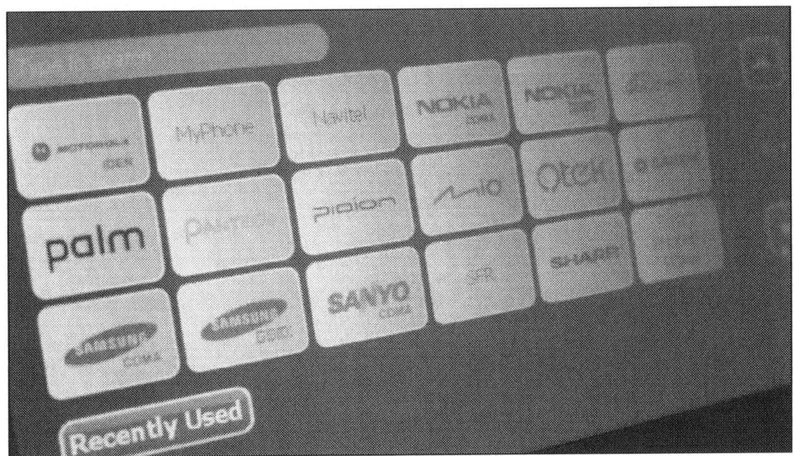

The UFED touch—vendor list screen

3. In the model menu, select the model of the device. Select **Physical Extraction**.
4. Select the location where you want to save the extraction—removable drive or the forensic workstation.
5. Follow the instructions exactly as listed on UFED Touch. Make sure you use the exact cable, and remove the battery when prompted.
6. The phone will enter download mode and display a logo. Next, connect the phone to UFED Touch and press continue.
7. Connect the external drive (to save the extracted data) to the target port of UFED Touch.
8. This will prompt UFED Touch to automatically move to the extraction screen. At this stage, you might be prompted to perform some of the phone connection steps. Do so if prompted.

9. Once the process is complete, the extracted data can be viewed and analyzed using the UFED Physical Analyzer application as shown in the following screenshot:

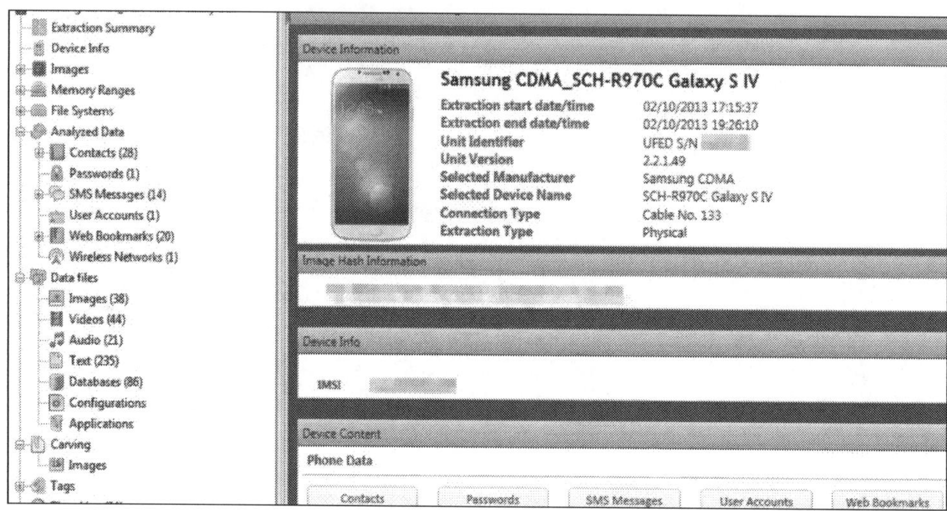

The UFED Physical Analyzer application

MOBILedit

As per the vendor, the **MOBILedit** forensic tool can be used to view, search, or retrieve data from a phone, including call history, phonebook, text messages, multimedia messages, files, calendars, notes, reminders, and application data such as Skype, Dropbox, Evernote, and more. It will also retrieve phone information such as IMEI, operating systems, firmware including SIM details (IMSI), ICCID, and location area information. Depending on the circumstances, MOBILedit is also able to retrieve deleted data from phones and bypass the passcode, PIN, and phone backup encryption. The setup file can be downloaded from `www.mobiledit.com` and can be installed easily. Once installed, perform the following steps to extract information from an Android phone using the MOBILedit software:

1. Ensure that USB debugging is enabled on the device and connect the Android device to the forensic workstation using a USB cable. MOBILedit attempts to detect the device, and to install the **Connector** app on the device, as shown in the following screenshot:

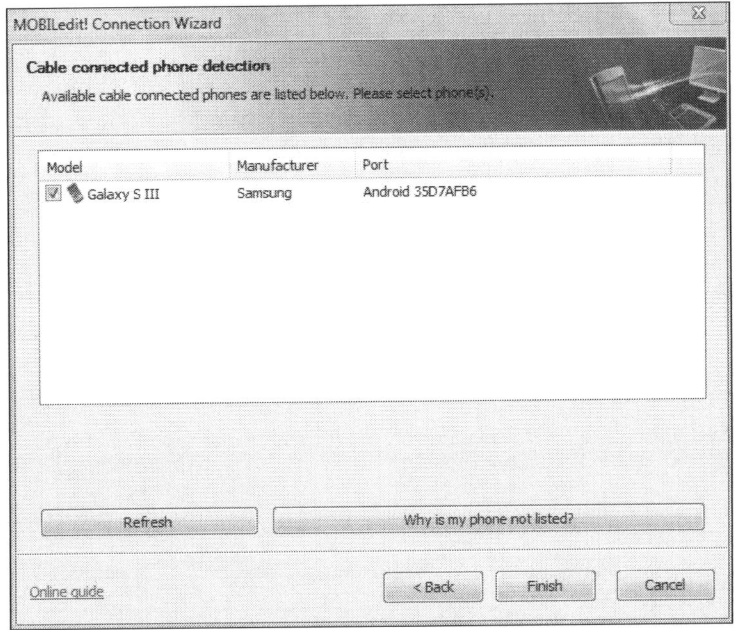

The MOBILedit connection wizard

2. MOBILedit then presents you with options to back up certain data. Once this is done, the tool displays statistics and the application data that can be used for analysis, as shown in the following screenshot:

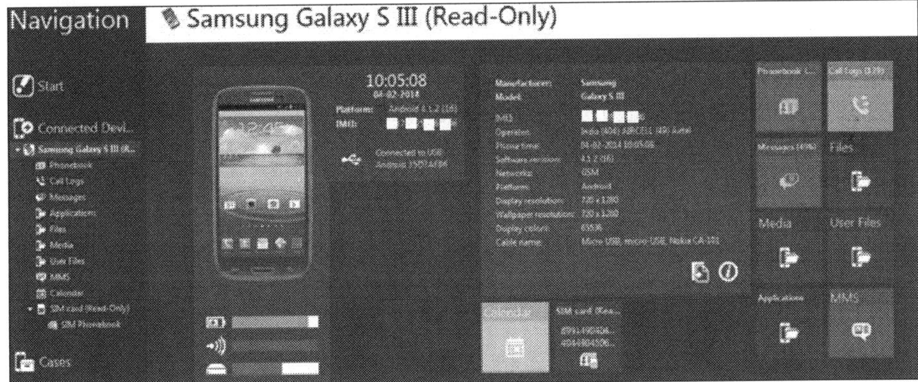

The MOBILedit tool results

3. Under the **Navigation** tab, click on any item to view the results. For instance, click on the **Phonebook** link to view all the contacts stored within the phone book including phone numbers, e-mail addresses, and more. Similarly, you can view the information about call logs by clicking on the **Call Logs** option, as shown in the following screenshot:

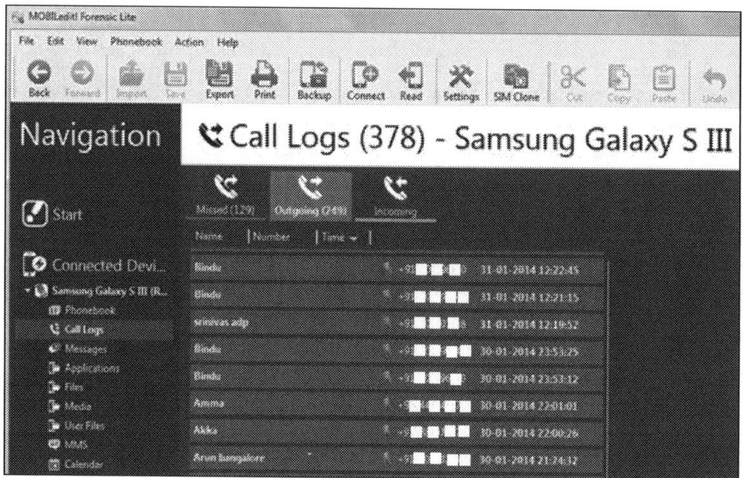

The MOBILedit tool—Call logs option

MMS, calendar, files on the SD card, and more, can be viewed by navigating through the available options.

Autopsy

Should manual examination or file carving be required, it is best to use a forensic tool that provides access to the raw files on the Android device. **Autopsy**, the GUI-based upon the **Sleuth Kit**, runs on a Windows forensic workstation and can be downloaded from `http://www.sleuthkit.org/autopsy/`. Autopsy currently provides analytical support for Android devices. Both open source and Law Enforcement modules are available for Autopsy. These modules provide additional file carving and parsing support for applications and files found on Android devices and SD cards. For example, the open mobile forensics module provides mobile device parsing capabilities to pull out artifacts such as calls, SMS, chats, pictures, and more.

Analyzing an Android in Autopsy

In this example, we will be using a physical image of the Samsung Galaxy SIII. This device was physically extracted using Cellebrite UFED Touch. The following steps should be performed to correctly mount an Android image and to start your examination:

1. Download and install the current version of Autopsy from `www.thesleuthkit.org`.
2. Launch Autopsy and select the option to create a new case as shown in the following screenshot:

The Autopsy tool screen

3. Fill out the case information and click on **Finish**.

4. Select **Image File** and navigate to the physical image of the Android device as shown in the following screenshot. If more than one image file is provided for the Android, simply select the first one.

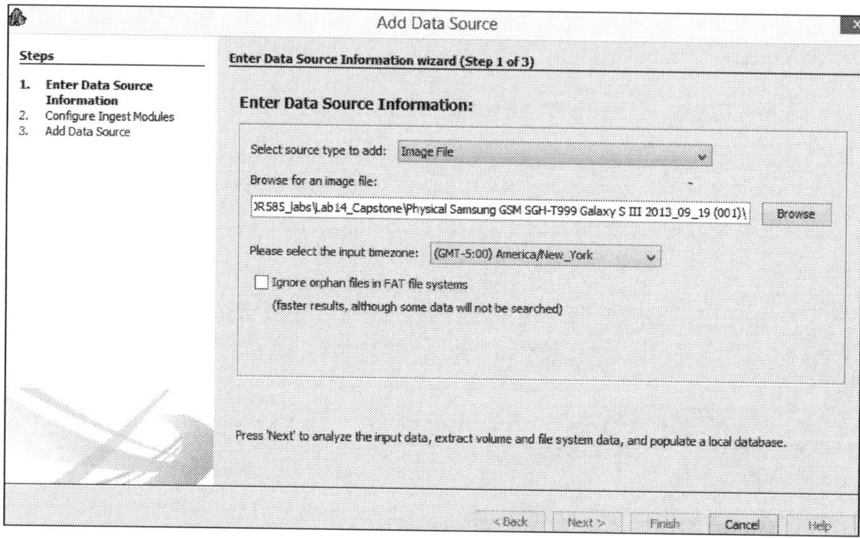

Autopsy image loading

5. Select the ingest modules you wish to run against the Android device. The module selections are shown in the following screenshot. Note that Law Enforcement modules are not listed and are provided only to those working in Law Enforcement and the Federal Government. The following screenshot shows the ingest modules:

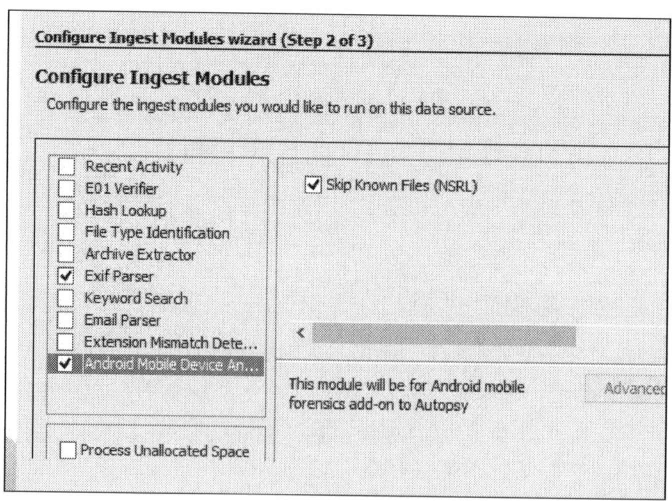

Autopsy ingest modules

6. Select **Next** and **Finish**, and Autopsy will begin to parse and load the Android image file. Unlike other forensic tools, Autopsy provides results as quickly as they are recovered to save the preprocessing time and allow the examiner direct access to the data involved in their investigation. The results appear as shown in the following screenshot:

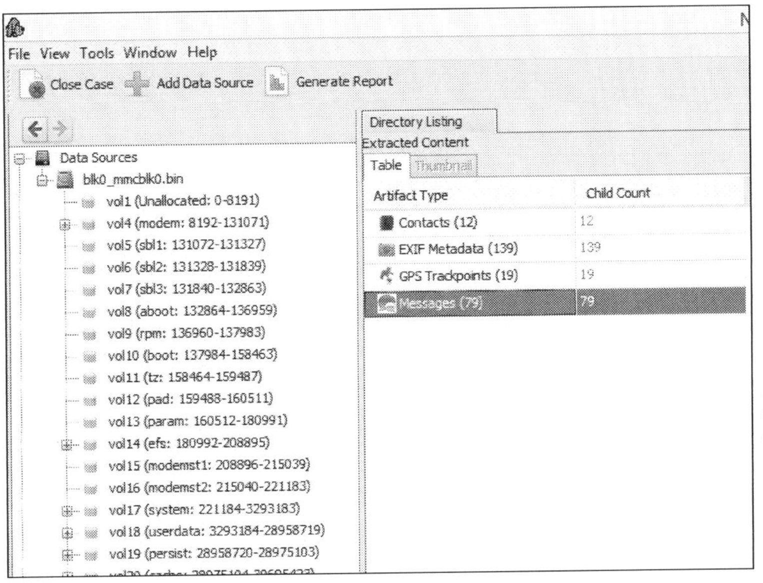

Autopsy results

Summary

Reverse engineering Android apps is the process of retrieving source code from an APK file. By using certain tools such as **dex2jar**, Android apps can be reverse engineered in order to understand the functionality of the app and data storage, define malware, and more. Forensic tools, such as AFLogical, Cellebrite, MOBILedit, and Autopsy, are just a few of the tools that are helpful to an examiner. They not only save time but also effort. A step-by-step explanation of using these was covered in this chapter. Unlike Android devices, data stored on Windows Mobile devices is difficult to extract and analyze. The next chapter provides a glance at performing forensics on Windows Mobile devices.

12
Windows Phone Forensics

Windows mobile devices are becoming more widely used and may be encountered during forensic investigations. Locating and interpreting digital evidence present on these devices requires specialized knowledge of the Windows Phone operating system and may not always be possible. Commercial forensic and open source tools provide limited support for acquiring user data from Windows devices. As Windows mobile devices are relatively new, most forensic practitioners are unfamiliar with the data formats, embedded databases used, and so on. This chapter provides an overview of Windows Phone forensics, describing various methods of acquiring and examining data on Windows mobile devices.

Windows Phone OS

Windows Phone is a proprietary mobile operating system developed by Microsoft. It was launched as a successor to Windows Mobile, but does not provide backward compatibility with the earlier platform. Windows Phone was first launched in October 2010 with Windows Phone 7. The version history of the Windows Phone operating system then continued with the release of Windows Phone 7.5, Windows Phone 7.8, and Windows Phone 8. Although the market share of this operating system is limited, there is certainly a case for optimism based on the following two reasons:

- The computer operating system market is still heavily dominated by Windows. This gives Windows Phone OS greater flexibility to provide users with a computer environment with which they are familiar.
- Microsoft's decision to acquire Nokia could be a significant factor in improving its market share of mobile operating systems.

The following sections will describe more about Windows Phone 7, its features, and the underlying security model. We believe the data is stored similarly on Windows Phone 8, so the methods defined in the following sections should work on both operating systems.

Unlike Android and iOS, Windows Phone comes with a new interface, which uses so-called **tiles** for apps instead of icons, as shown in the following figure. These tiles can be designed and updated by the user. Similar to other mobile platforms, Windows Phone allows for the installation of third-party apps. The apps can be downloaded from Windows Phone Marketplace, which is managed by Microsoft.

The Windows Phone home screen

Security model

The security model of Windows Phone is designed to make sure that the user data present on the device is safe and secure. The following sections are a brief explanation of the concepts on which Windows Phone security is built.

Windows chambers

The Windows Phone OS 7.0 is heavily built on the principles of least privilege and isolation. To achieve this, Windows Phone introduced the concept of **chambers**. Each chamber has an isolation boundary within which a process can run. Depending on the security policy of a specific chamber, a process running in that chamber has the privilege to access the OS resources and capabilities (`https://www.msec.be/mobcom/ws2013/presentations/david_hernie.pdf`). There are four types of security chambers. The following is a brief description of each one of them:

- **Trusted Computing Base (TCB)**: Processes here have unrestricted access to most of the Windows Phone 7 resources. This chamber has the privilege to modify policies and enforce the security model. The kernel runs in this chamber.
- **Elevated Rights Chamber (ERC)**: This chamber is less privileged than the TCB chamber. It has the privileges to access all resources except the security policy. This chamber is mainly used for services and user-mode drivers, which provide functionality intended for use by other applications on the phone.
- **Standard Rights Chamber (SRC)**: This is the default chamber for preinstalled applications, such as Microsoft Outlook Mobile 2010.
- **Least Privileged Chamber (LPC)**: This is the default chamber for all the applications that are downloaded and installed through the Marketplace Hub (also known as the Windows Phone Marketplace).

Capability-based model

Capabilities are defined as the resources on the phone (camera, location information, microphone, and more), which are associated with security, privacy, and cost. The LPC has a minimal set of access rights by default. However, this can be expanded by requesting more capabilities during the installation. Capabilities are granted during the app installation and cannot be modified or elevated during runtime.

To install an app on a Windows phone, you need to sign in to Marketplace with a Windows Live ID. During installation, apps are required to ask the user for permission before using certain capabilities, an example of which is shown in the following screenshot:

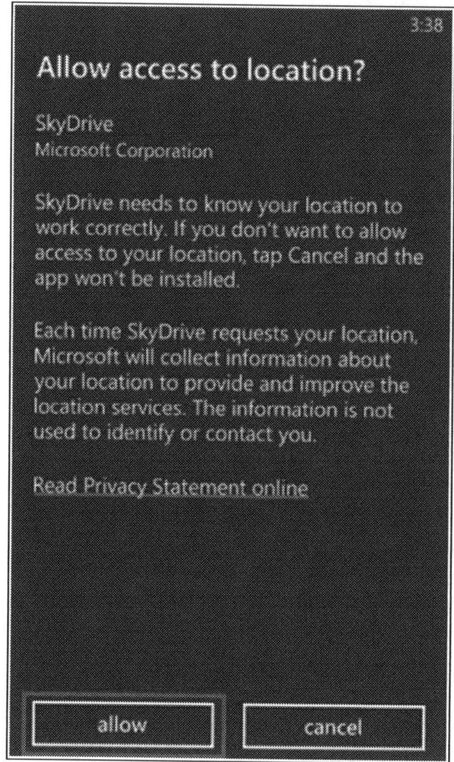

Windows app requesting user permissions

This is similar to the permission model in Android. This gives the user the freedom to learn about all the capabilities that an application has before installing the application. The list of all capabilities is included in the application manifest file `WMAppManifest.xml`, which can be accessed through visual studio or other methods defined at `http://developer.nokia.com/community/wiki/How_to_access_Application_Manifest_%28WMAppManifest.xml%29_file_at_runtime`.

App sandboxing

Apps in Windows Phone run in a sandboxed environment. This means every application on Windows Phone 7 runs in its own chamber. Applications are isolated from each other and cannot access the data of other applications. If any app needs to save information to the device, it can do so using the isolated storage, which is restricted from access by other applications. Also, the third-party applications installed on Windows Phone cannot run in the background, that is, when the user switches to a different application, the previously used application is shut down (although the application state is preserved). This ensures that the application cannot perform activities such as communicating over the Internet when the user is not using the application. These restrictions also make the Windows Phone less susceptible to malware.

Windows Phone file system

The Windows Phone 7 file system is more or less similar to the file systems used in Windows XP, Windows Vista, or Windows 7. From the root directory, one can reach different files and folders available on the device. From a forensic perspective, the following are some of the folders that can yield valuable data. All the mentioned directories are located in the root directory.

- **Application Data**: This directory contains data of preinstalled apps on the phone such as Outlook, Maps, and Internet Explorer.
- **Applications**: This directory contains the apps installed by the user. The isolated storage, which is allocated or used by each app, is also located in this folder.
- **My Documents**: This directory holds different Office documents such as Word, Excel, or PowerPoint. The directory also includes configuration files and multimedia files, such as music or videos.
- **Windows**: This directory contains files related to the Windows Phone 7 operating system.

Windows Phone also maintains **Windows registry**, a database that stores environment variables on the operating system. The Windows registry is basically a directory that stores settings and options for the Microsoft operating system.

Data acquisition

Acquiring data from a Windows Phone is challenging for forensic examiners, as physical and logical methods defined in previous chapters are not commonly supported. One of the most common techniques in data acquisition is to install an application or agent on the device, which extracts as much data as possible from the device. This could result in certain changes on the device but nevertheless, it is still forensically sound if the examiner follows standard protocols. These protocols include proper testing to ensure no user data is changed, validation of the method on a test device, and documenting all steps taken during the acquisition process. For this acquisition method to work, the app needs to be installed with the privileges of Standard Rights Chamber. This may require the examiner to copy the manufacturer's DLLs, which have higher privileges into the user app. This allows the app to access methods and resources that are usually limited to native apps.

Most examiners rely on forensic tools and methods to acquire mobile devices. Again, these practices are not readily available for Windows Mobile devices. Keep in mind that to deploy and run an app on Windows Phone, both the phone and the developer must be registered and unlocked by Microsoft. This restriction can be bypassed by unlocking the device using tools such as **ChevronWP7**. This tool basically allows the bypassing of Marketplace procedure and allows you to sideload (run unsigned applications without the restrictions listed) an unpublished application.

Sideloading using ChevronWP7

As explained earlier, in order to install the app that provides access to the file system of the phone, we first need to unlock the device (similar to jailbreaking on iOS devices). This method will only work on a Windows Phone that is not locked with a passcode. This can be done using the ChevronWP7 tool by performing the following steps:

1. Download `ChevronWP7.exe` and `ChevronWP7.cer` files. Note that these files are often removed and are not always available on the same site. One location that currently has the files available for download is `http://www.4shared.com/file/HQGmwIRx/ChevronWP7.htm?locale=en`.
2. Install `ChevronWP7.cer` on the Windows Phone. Note that the methods for installing ChevronWP7 may require techniques not standard to forensic practices. Thus, all methods must be tested on a sample Windows Phone to ensure user data is not lost in the process of attempting to extract the data. One method for installing ChevronWP7 includes sending it to an e-mail and accessing it. This method should be used as a last resort when all other acquisition methods fail.

3. Connect the phone to your computer and make sure that the device is not passcode-locked. If the device is locked and the password is known, enter the password only when prompted by the computer. *Do not* guess the password on the Windows Phone as multiple incorrect guesses may wipe the user data.

4. Run `ChevronWP7.exe` and check both the boxes shown in the following screenshot and click on **Unlock**. This enables the *developer unlock* on the device and also enables you to install any third-party app without a Marketplace developer account.

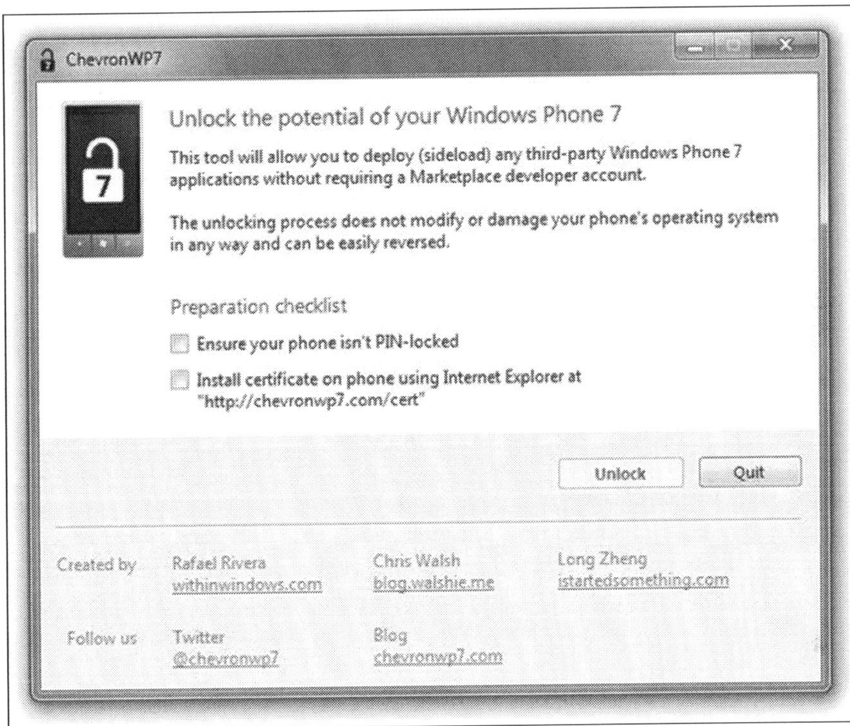

The ChevronWP7 tool

To execute native code in a user app, the `Windows.Phone.interopService` DLL is used. This DLL provides the method `RegisterComDLL`, which can import native manufacturer DLLs. Hence by including this DLL in a user app, it is possible to execute native code within the app and get access to the entire file system of the phone, including the isolated storage.

Extracting the data

On an unlocked device (again, similar to a jailbroken iOS device), it is possible to run an app that can extract the user data present in the phone. The app **TouchXperience**, which comes along with the **Windows Mobile Device Manager (WPDM)**, can be used for this purpose. Windows Mobile Device Manager is the management software for Windows Phone 7. The client app TouchXperience extracts data such as the file system from the mobile device, and WPDM retrieves this data and converts it into a human readable graphical format. The following are the steps which will help a forensic examiner extract user data present on an unlocked Windows Phone device:

1. Download Windows Phone SDK 7.1 and the Zune software on the forensic workstation and install it (`http://www.microsoft.com/en-us/download/details.aspx?id=27570`).
2. Download the Windows Phone Device Manager on the workstation, and launch `WPDeviceManager.exe` (`http://touchxperience.com/windows-phone-device-manager/`).
3. Connect the device to the workstation, and it should be detected automatically. If it is not detected, make sure a passcode is not set on the device. If it is, this process may fail if the passcode is unknown.
4. Windows Phone Device Manager will automatically install the TouchXperience app when the phone is connected for the first time. Make sure you set what the software is allowed to do on the device (that is, make sure not to change the user data, not update date/time settings, or anything else that will modify the user data). Make sure to document that TouchXperience was installed in order to extract data from the Windows Phone as standard forensic methods provide little support for these devices.
5. Thereafter, the following screen is presented, which provides access to a vast amount of files present on the device:

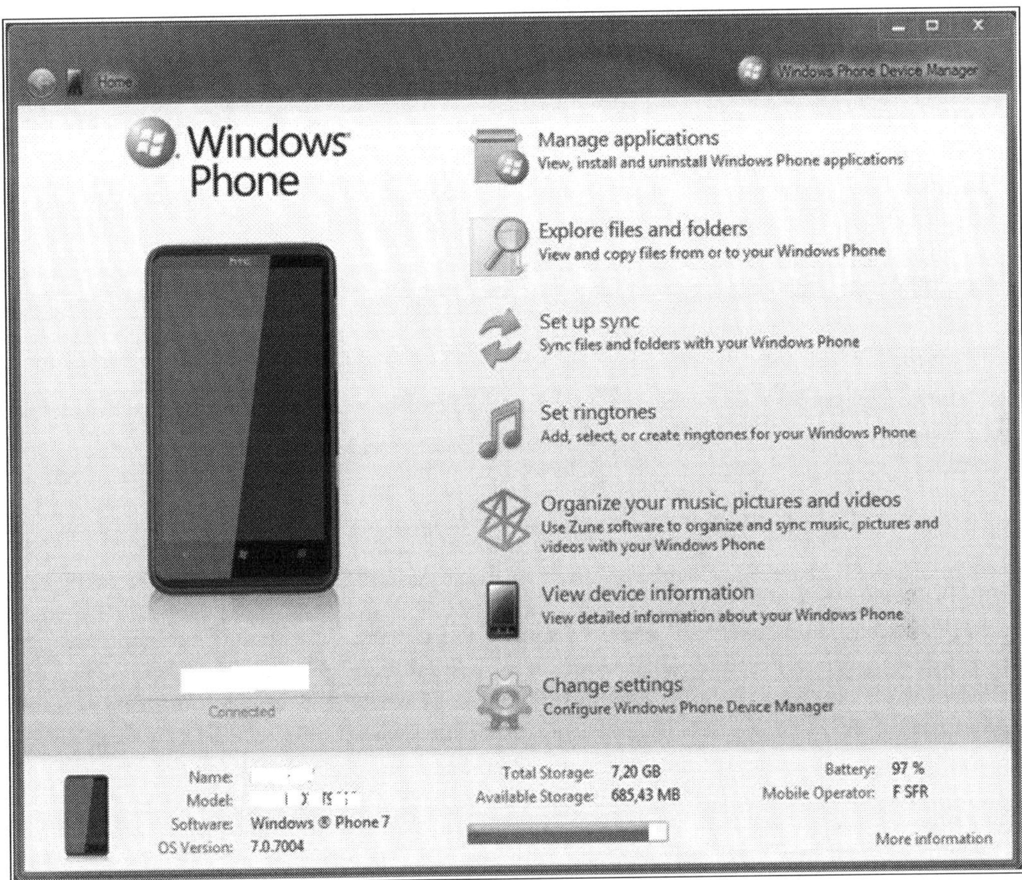

Windows Phone Device Manager

The home screen displays information about the model of the phone, OS version, and more. Click on **Manage applications** to see the information about installed apps on the device, as shown in the next screenshot. WPDM also provides other functionality, such as media management, synchronization of files and folders, and more. From a forensic point of view, the File Explorer is the most interesting part of this software. It provides read, write, and executable access to most of the files present on the Windows Phone 7 device.

Have a look at the following screenshot:

Name	Publisher	Installed On	Size	Version
Installed Applications				
TouchXplorer	Julien Schapman	28/02/2011	664,91 KB	1.0.0.0
TouchXperience	Julien Schapman	28/02/2011	2,42 MB	1.0.2.0
Bluetooth	Julien Schapman	28/02/2011	587,02 KB	1.0.0.0
Config. avancée	Julien Schapman	28/02/2011	1,31 MB	1.1.0.1
Éditeur de registre	Julien Schapman	28/02/2011	1,29 MB	1.1.0.0
Purchased Applications				
Config Connexion	HTC Corporation		913,10 KB	1.0.0.0
Convertisseur	HTC Corporation		1,82 MB	1.0.0.0
HTC Hub	HTC Corporation		18,04 MB	1.0.0.0

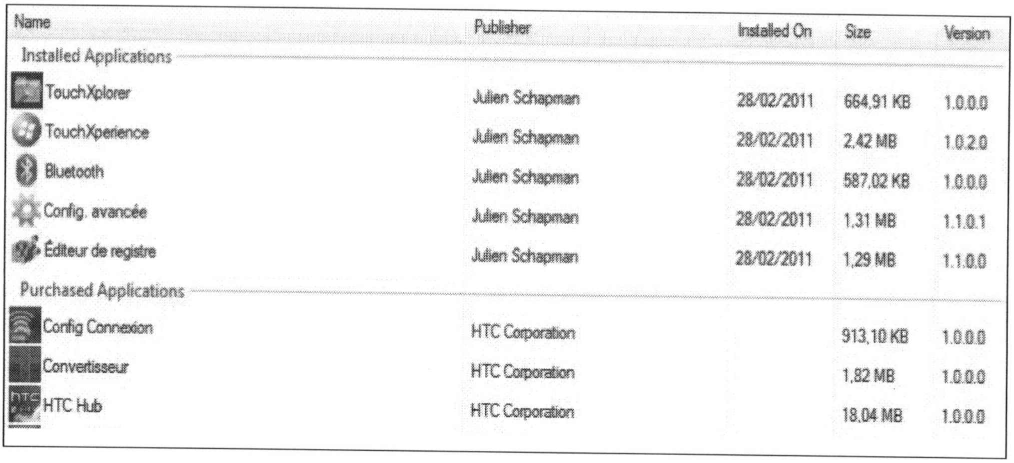

Windows Phone Device Manager—The Manage Applications screen

Using this acquisition technique, you can acquire two types of data: system data and application data. System data is mainly the data that is required to run the phone, and application data is the data created and used by different applications installed on the device. While system data may not contain data relevant to your investigation, application data is very much valuable. Regardless, all data should be acquired from any smartphone as the examination must be complete and capture all data contained on the device when possible. The following sections discuss the steps to be followed to extract application data from a Windows Phone device. The application data will contain the bulk of the user-created data and will provide the most value to your investigation.

Extracting SMS

All the incoming and outgoing short messages (SMSes) in Windows Phone 7 are stored in the file named `store.vol`, which is present under the directory `\Application Data\Microsoft\Outlook\Stores\DeviceStore`, as shown in the next screenshot. However, it is not possible to copy this file directly because this file is always in use. When the file is renamed (say `store.vol.txt` or `store.bkp`), it automatically creates a copy of the file. Once the copy is made, this file can now be examined using a normal text editor. Note that this file can also exist in the `\APPDATA\Local\Unistore` directory. Have a look at the following screenshot:

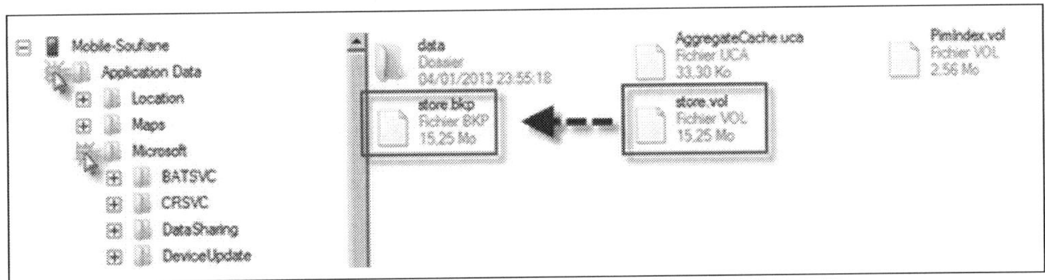

The store.vol file in Windows Phone

Extracting e-mail

Windows Phone 7 devices use Outlook as their standard e-mail client. This can be used to synchronize with various e-mail services such as Google, YahooMail, and more. Any data that belongs to Outlook is stored under the directory `\Application Data\Microsoft\Outlook\Stores\DeviceStore\data`, as shown in the following screenshot:

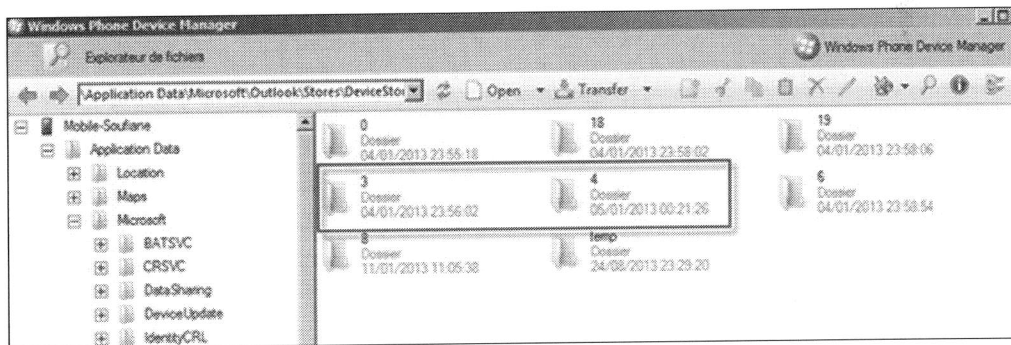

Windows Phone: extracting e-mail

As shown in the next screenshot, there are different folders present that contain different data. For example, folder `3` contains pictures of the user's contacts (e-mail receivers). This folder is being used as an example. This folder will not be consistently named folder `3` across Windows Phone devices. Have a look at the following screenshot:

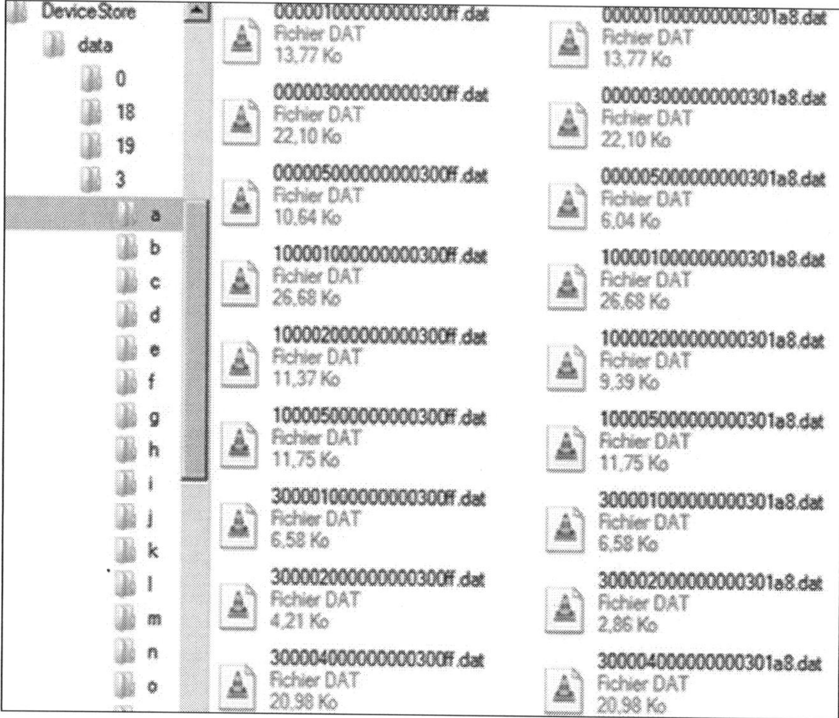

Windows Phone: folder 3

Although the files are present with the `.dat` extension, by renaming them to `.jpg`, we can view the pictures as shown in the following screenshot:

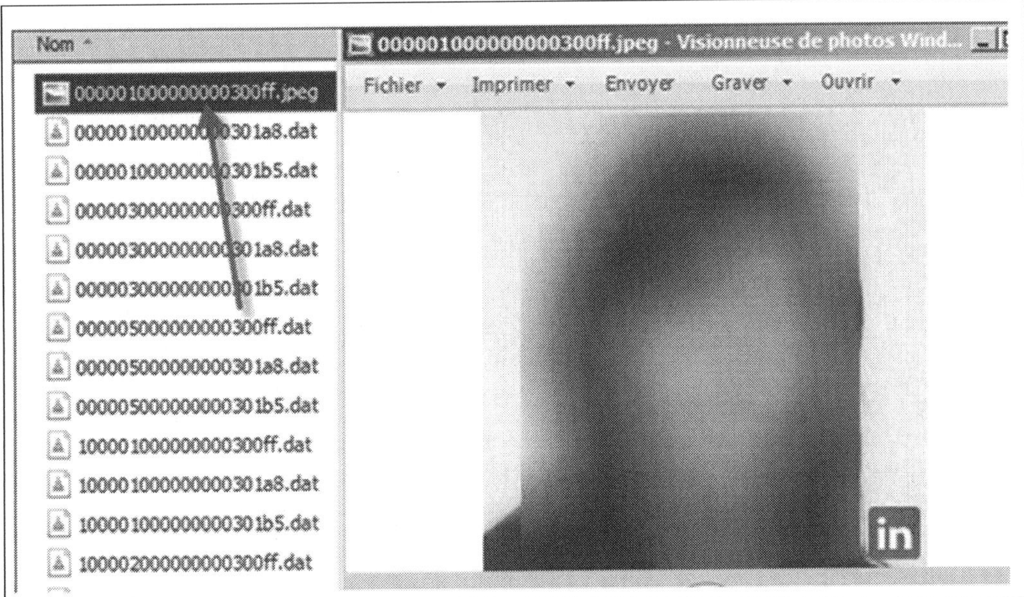

Windows Phone: renaming data files to JPG files

Similarly, folder `4` contains information about e-mail messages. By renaming the files to `HTML`, we can view the content of the e-mail messages. Again, each folder should be examined for relevance as they may contain e-mail messages, attachments, contacts, and more.

Extracting application data

The `Applications` folder contains all the applications installed on the phone. Each application has its own directory, which is identified with a unique application ID. Inside the application ID folder, there are other important folders, such as `Cookies`, `History`, `IsolatedStore`, and more. Most of the crucial information is usually present in the `IsolatedStore` folder. For example, as shown in the next screenshot, the `IsolatedStore` folder in Facebook contains the following data:

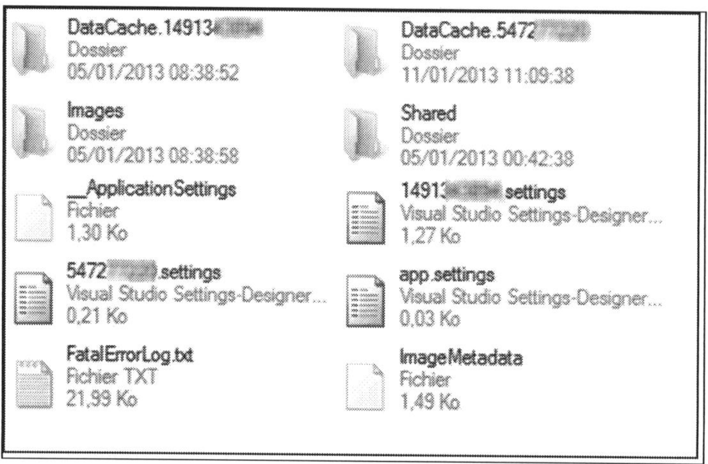

Contents of the IsolatedStore folder

By analyzing these folders, a forensic analyst can gather a lot of information that could aid in the investigation. The following are some of the findings from our Facebook app analysis example:

- The file `userid.settings` shown in the following screenshot contains the user's profile name and a link to the user's profile and profile picture.
- All the pictures used by the Facebook app are stored in the `Images` folder present in the directory `IsolatedStore`. To view the images, change the extension of the files to `JPG`.
- The `DataCache.userID` folder contains most of the information about the Facebook account. By parsing this folder, information about friends, friend requests, messages, and more can be obtained. This is straightforward as all the files, once extracted, can be manually examined for relevance to the investigation.

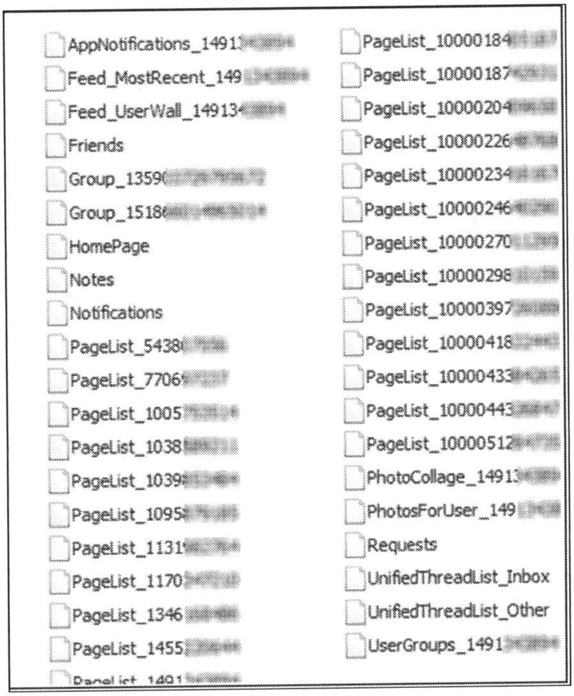

The DataCache.UserID folder of the Facebook app

Similarly, by examining the Internet Explorer app, a forensic examiner can gather information about the sites visited by the user. All this data can be found under the `Application Data\Microsoft\Internet Explorer` folder. By analyzing the Maps application, information about the user location and other details can be obtained. The call logs can be recovered from `\APPDATA\Local\UserData\Phone` on most devices. Keep in mind that the location may vary depending on the OS and the Windows device. However, the directory containing the data (`phone`, `store.vol`, and so on) remains the same. A great source for conducting forensics on a Windows Phone device can be found at `http://cheeky4n6monkey.blogspot.com/2014/06/monkeying-around-with-windows-phone-80.html`.

Summary

Acquiring data from Windows Phone devices is challenging as they are secure, and as commercial forensic tools and open source methods do not provide easy solutions for forensic examiners. chip-off, JTAG, and the methods defined in this book are some of the methods that provide access to user data on Windows Phone devices. The biggest challenge is getting access to the device, acquiring the data, and extracting the raw files for analysis. Once the data is available, all the information about SMS, e-mail, application data, and more can be analyzed by the examiner. Again, the device must not contain a passcode, must be unlocked (jailbroken/rooted), and will be modified by the examiner in order to extract the data using the methods defined in this chapter. While some may challenge us and say these methods are not common in forensic practices, they must realize that these methods may be the only way to obtain user data from Windows Phone devices. In the next chapter, we will cover BlackBerry forensics, which, while challenging, is more supported by commercial and open source methods.

13
BlackBerry Forensics

BlackBerry devices come with the **Research in Motion** (**RIM**) software implementation of proprietary wireless protocols. BlackBerry devices pose a significant challenge to forensic examinations due to the lack of physical parsing support and device encryption. This chapter will cover the various security features that come with BlackBerry devices, the available techniques to extract data from a device, and the best methods to analyze the data extracted.

BlackBerry OS

BlackBerry OS is a proprietary mobile operating system developed by the Canadian company RIM used on all BlackBerry devices until BlackBerry 10, which introduced QNX. BlackBerry RIM is now referred to as BlackBerry Limited. The initial BlackBerry operating system is known to support specialized functions, such as trackball, trackwheel, trackpad, and more. BlackBerry OS was initially released in 1999 for the device Pager BlackBerry 580. BlackBerry QNX (OS 10) uses a Linux variant that was initially introduced with the BlackBerry Playbook and is now used on BlackBerry devices. With QNX, BlackBerry World and Balance were introduced along with other features more comparable to Android and iPhone (`http://searchitchannel.techtarget.com/feature/Introduction-to-the-BlackBerry`).

The following table provides information about the version history of BlackBerry OS:

Version	Release year
1	1999
3.6	2002
5	2008
6	2010
7	2011
7.1	2012
10	2013
10.1	2013
10.2	2013

BlackBerry OS versions

The BlackBerry OS offers native support for corporate mail through MIDP, which allows wireless syncing with Microsoft Exchange, Lotus Domino and e-mail, contacts, calendar, notes, and more, while used along with the BlackBerry Enterprise Server. This OS additionally supports WAP 1.2. With the advent of Android and iOS, the market share of BlackBerry OS has steadily decreased over the years. Nevertheless there are more than 70 million BlackBerry users worldwide and these devices are frequently encountered during forensic investigations, especially for internal corporate investigations. The **BlackBerry Enterprise Server** (**BES**) consists of software that facilitates corporate messaging to allow the syncing of corporate e-mail with the user's device. A BES administrator of an IT department normally manages BES services. The **BlackBerry Internet Service** (**BIS**) is a service that allows the user to configure up to 10 e-mail accounts to sync to the BlackBerry device.

BlackBerry allows the installation of third-party apps from BlackBerry World, which is the app distribution service. BlackBerry apps are developed using a **Java Development Environment** (**JDE**) or RIM's **Mobile Data System (MDS)**. If the application can run independently of a BlackBerry solution, such as BIS or BES, a Java application would serve the purpose. If the application requires e-mail for functionality or needs support from a BlackBerry device to help it operate, MDS is usually preferred to develop the application.

Security features

There are two types of BlackBerry users—consumers who buy and use the device, and enterprise users who are provided with the BlackBerry device by their employers. The consumer devices are usually configured to use the BIS, whereas the enterprise user devices are configured to use BES. In a BES environment, security is usually enforced by the enterprise through appropriate settings and application controls.

Although BlackBerry uses a proprietary operating system, its third-party application framework is mostly based on Java. Third-party apps that are not signed have very limited access to this restrictive functionality. Even in the case of signed applications, user permission is needed to perform important actions such as calling a number, accessing a contact, and more. BlackBerry apps are written in Java and then compiled into **COD** files. But before compiling the apps, they are preverified for certain security checks and are tagged to confirm that the checks have been carried out. When the **Java Virtual Machine (JVM)** present on BlackBerry loads the class, it can cross-check and perform its own verification much faster. Any changes to the code after the preverification can be easily detected at runtime and JVM will prevent their execution. This makes BlackBerry a secure platform that is less susceptible to malware when compared to other smart devices.

In order for an application to get full access to all the APIs, the application must be signed by RIM. When the developers first register with RIM, they receive a developer key. Using the signing tool provided by RIM, the SHA1 hash of the application can be sent to RIM. Upon receiving this, RIM generates a signature, which is then sent back to the developer and added to the application. When the signed application is loaded onto a BlackBerry device, the JVM links the COD file with the API libraries and checks that the application has the required signatures. If the required signature is not present, JVM will refuse to link the application to the respective APIs, and hence, the application will fail at runtime. This way, BlackBerry ensures security for the device through the code-signing process.

The security strength of BlackBerry can be attributed to the granular control that it provides through the IT policies present on the BES. It is important to note that many of the security controls that are enabled with BES devices are not present in consumer devices that use BIS. BES devices come with various security features, as follows:

- **Data protection**: All the data that is sent between the BES and a BlackBerry device is encrypted using BlackBerry transport layer encryption. Before the BlackBerry device sends a message, it compresses and encrypts the message using the device transport key. When the BES receives a message from the BlackBerry device, the BlackBerry Dispatcher decrypts the message using the device transport key and then decompresses the message. The BlackBerry uses AES or Triple DES as the symmetric key cryptographic algorithm for encrypting data. By default, the BES uses the strongest algorithm that both the BES and BlackBerry devices support for the BlackBerry transport layer encryption. More information on data protection can be found at `http://btsc.webapps.blackberry.com/btsc/viewdocument.do;jsessionid=E8567E865DBC9668D3F8740BEB9D65E6?externalId=KB13160&sliceId=1&cmd=displayKC&docType=kc&noCount=true&ViewedDocsListHelper=com.kanisa.apps.common.BaseViewedDocsListHelperImpl`.
- **Protection of data and encryption keys on the device**: If the content protection option is turned on, BlackBerry devices can be configured to encrypt data stored on the device. By default, a locked BlackBerry device was created to use AES-256 encryption to encrypt stored data and an ECC public key to encrypt data that is sent to the locked BlackBerry device (`http://docs.blackberry.com/en/admin/deliverables/25763/Encrypting_user_data_on_a_locked_BB_device_834471_11.jsp`). Also, BlackBerry is designed to protect the encryption keys that are stored on the device. The device encrypts the encryption keys when the device is locked.
- **Better control over the device**: You can use an IT policy to control a BlackBerry device. The IT policy usually consists of multiple policy rules that manage the security and behavior of the BES. For example, using the IT policy rules, the following security features on a BlackBerry device can be controlled:
    - Encryption of data transmitted between the BlackBerry server and the device
    - Connections that use Bluetooth wireless technology
    - Protection of user data stored on the BlackBerry device
    - Control of protected device resources, such as the camera or GPS, that are available to third-party applications

  In addition to all this, the BES administrator can also reset user passwords for the BlackBerry device and initialize a remote wipe, which must be considered during forensic investigations.

BlackBerry security is a huge hurdle for forensic examiners. While a BES administrator can be used to reset a device password, which may allow an examiner to access the device, they can also remotely wipe the device. Thus, following steps similar to those for Android and iOS, the examiner must place the device in airplane mode and disable all remote connections to the device. A BlackBerry wipe initiated via the BES can exist for an extended period of time. This means that even if the battery is removed from the device and the BlackBerry boots, the wipe could immediately be sent to a connected BlackBerry. While Android and iOS proved to be easier to access when locked, a locked BlackBerry device is more difficult. The level of protection on these devices may render the extracted data encrypted even after a JTAG or chip-off extraction. Physical support, to include both acquisition and analysis, is limited for BlackBerry devices. As described in the following sections, most of the data is obtained by simply obtaining a backup of the device.

Data acquisition

While the sales of BlackBerry devices is on the decline, they are still encountered during forensic investigations. Commercial forensic tools provide limited support for BlackBerry devices in comparison to other smartphones. Even worse, open source methods are not available for data acquisition of BlackBerry devices. Hence it is important for the examiners to understand all possible methods of data extraction available for these devices. The following sections discuss the various steps involved in acquiring data from a BlackBerry device.

Standard acquisition methods

Standard forensic acquisition methods can be applied to BlackBerry devices. However, encrypted and locked devices may not be possible to acquire, and it will be even more difficult (if not impossible) to analyze whether the password or encryption keys are present. The level of acquisition support available depends on the forensic kit, the device model, and the security level currently being used on the BlackBerry device. As explained in previous chapters, logical and physical (to include file system) acquisition methods are possible on BlackBerry devices. The Cellebrite UFED Touch provides the greatest level of physical acquisition support for BlackBerry devices (at the time of writing this). The following two images show the different support provided by the Cellebrite UFED Touch on two different models of BlackBerry.

Note that one model has full acquisition support while the other only offers logical acquisition.

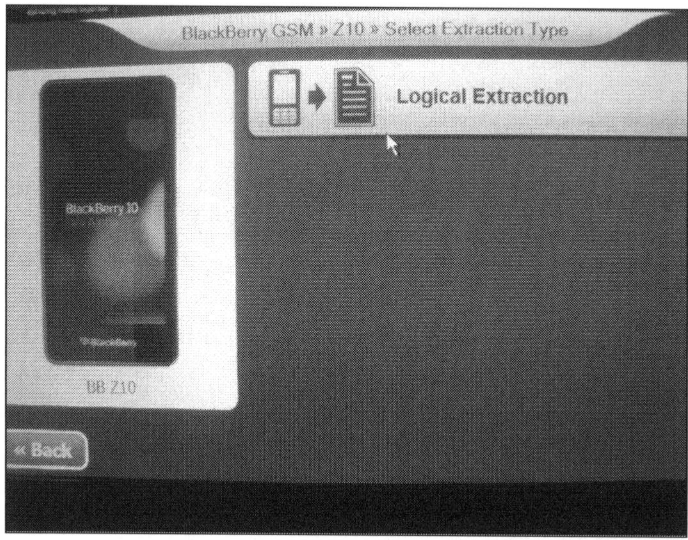

The BlackBerry Z10 support in Cellebrite UFED Touch

The following image shows that the BlackBerry Z10 device can only be logically acquired using the Cellebrite UFED Touch. When attempting to acquire a BlackBerry 8300 using the UFED Touch, logical, physical, and file system acquisition support is possible, as shown in the following image:

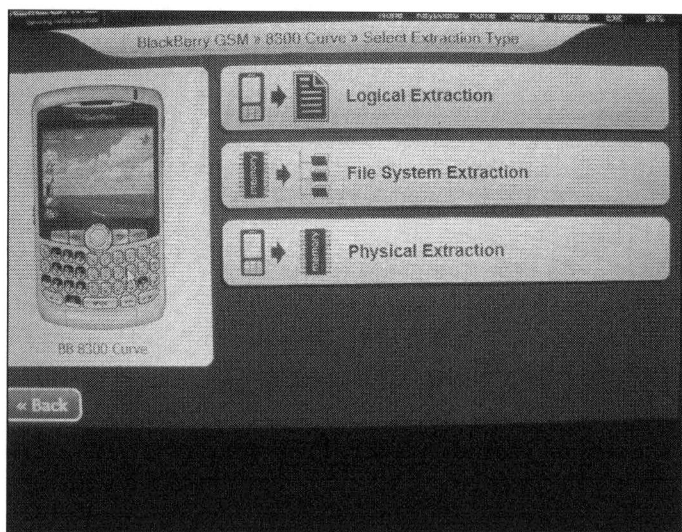

The BlackBerry Curve support in Cellebrite UFED Touch

The device passcode must be known for physical acquisition. This is one of the major differences between BlackBerry's physical acquisition and Android and iOS. Keep in mind even if a BlackBerry device is physically acquired, any tool currently available to forensic examiners may not support the analysis portion. These challenges will be discussed in the analysis section. Logical support for BlackBerry devices is more common and is supported by most commercial forensic tools to include Oxygen Forensics, Microsystemation XRY, Cellebrite UFED Touch, and more. Most BlackBerry support provided by commercial forensic tools applies to devices using BlackBerry OS (Java-based) and not QNX (BlackBerry 10 OS).

A physical acquisition of a BlackBerry device will capture a complete binary image of the BlackBerry device. This method of acquisition normally requires the BlackBerry to be powered off and intercepts the data prior to the device booting. File system acquisitions may be possible using commercial tools if the device passcode is known. This method of acquisition normally captures data from the device and the SD card. As mentioned, even if a physical or file system acquisition is supported and successful, the examiner should always obtain a logical acquisition to avoid situations where physical data parsing is not supported by the forensic analysis tool. One of the biggest errors in BlackBerry forensics occurs when an examiner obtains only a physical image, returns the device to the user/suspect, and then realizes the data is encrypted or cannot be parsed by their analytical tool. Make sure you do not find yourself in this position by taking the time to acquire the device using all possible methods. The following screenshot shows security prompts that the examiner may encounter during the acquisition and/or analysis of a BlackBerry device:

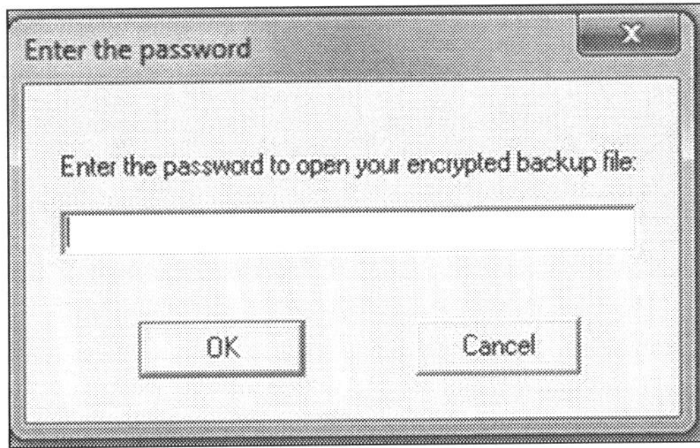

The encrypted backup file password prompt

The preceding screenshot shows the prompt for the user to enter the password for the encrypted backup file when attempting to open the image in Cellebrite Physical Analyzer. All forensic tools that attempt to parse the image or backup file for analysis will require the password. Without the password, the examiner cannot access the image.

The following screenshot shows the prompt to open the image file in Oxygen Forensics Suite.

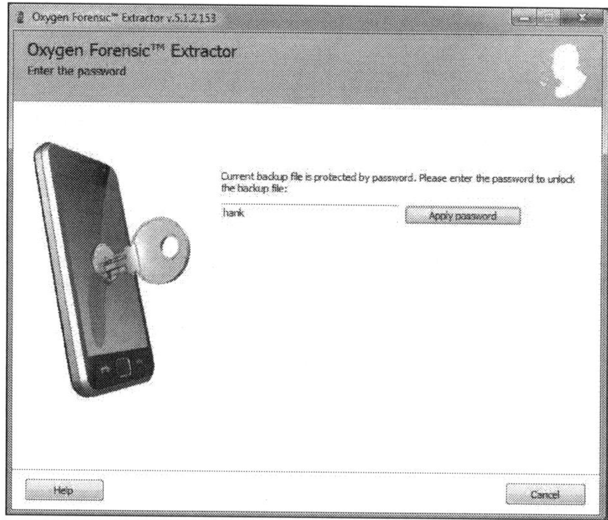

The encrypted backup file password prompt in Oxygen Forensics Suite

Creating a BlackBerry backup

With BlackBerry devices, a significant amount of data can be extracted using the **BlackBerry Desktop Manager (BDM)** or **BlackBerry Link** (BlackBerry 10 devices), which can be downloaded for free. This method of acquiring data from a BlackBerry device sometimes proves to obtain and provide data for examiners to analyze. Again, the passcode must be known for the examiner to create a backup of a BlackBerry device. Acquiring this logical backup is recommended because it can provide a form of validation for the data acquired through forensic tools. The backup file exists as a BBB or IPD file and contains different types of data stored on the BlackBerry device, including call logs, calendar items, contacts, pictures, e-mail, and more.

A **BlackBerry Backup (BBB)** file is created when BDM v7.0 and later versions or a Mac computer is used to create the backup file. The BBB file will either be a ZIP container comprised of an IPD file or DAT files, depending on the method to create the backup file. A BBB file that contains an IPD file has the same file header as a ZIP file. In Hex, this file header is `0x504B`. An **Inter@ctive Pager Backup (IPD)** is created when BDM v6.0 or earlier is used to create the backup file. Commercial forensic tools may also create BlackBerry backup files and use the IPD format. Shafik Punja maintains a blog, highly dedicated to his work on BlackBerry, that provides a deeper look into BlackBerry backup files (`http://qubytelogic.blogspot.com/`).

It is important to note here that, by default, the **BDM** is configured to synchronize some data between the device and the computer. Hence, it is important to disable this feature in order to prevent any changes of data on the device. In a forensic process, even a minor change, such as altering the time zones on a device, would make it difficult for an investigator to analyze when specific events exactly occurred and will be even more difficult to defend in court. Hence it is necessary to disable the synchronization process in the BDM by disabling the options as shown in the following screenshot. The option **Update device data and time** is selected by default, so it is necessary to explicitly deselect this option. It is the examiner's job to ensure that total control is maintained during the entire forensic process. This means that the forensic workstation is sterile and free of old data and that the tools are not set to automatically read/write data to and from the BlackBerry device. If the BDM requires the device be connected in order to select the options, it is wise to attempt the settings with a test BlackBerry device of the same model as your evidence.

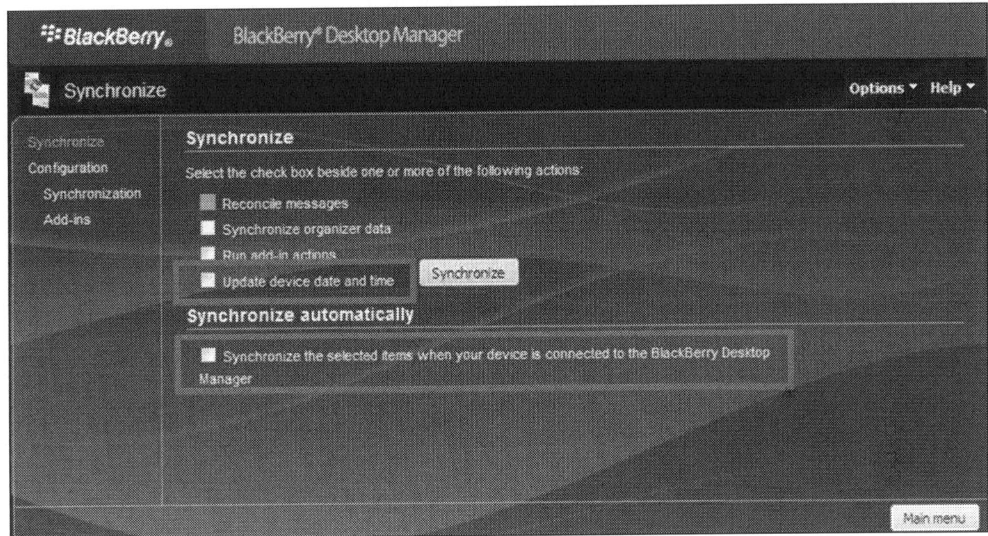

BlackBerry Desktop Manager

The following is the step-by-step procedure to create a backup of the BlackBerry device using **BlackBerry Desktop Manager**:

1. On the forensic workstation, install **BlackBerry Desktop Software**. Certain versions of BDM may be required to connect with older BlackBerry devices.

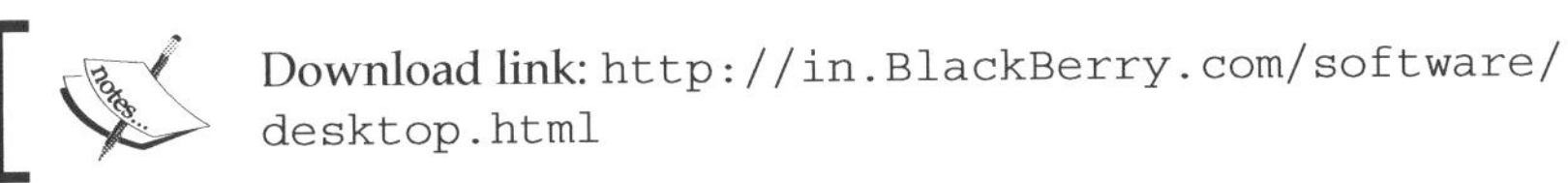

Download link: `http://in.BlackBerry.com/software/desktop.html`

2. Connect the BlackBerry device to the workstation and observe that the device is detected.
3. Click on **Back up** under **Device**, as shown in the following screenshot:

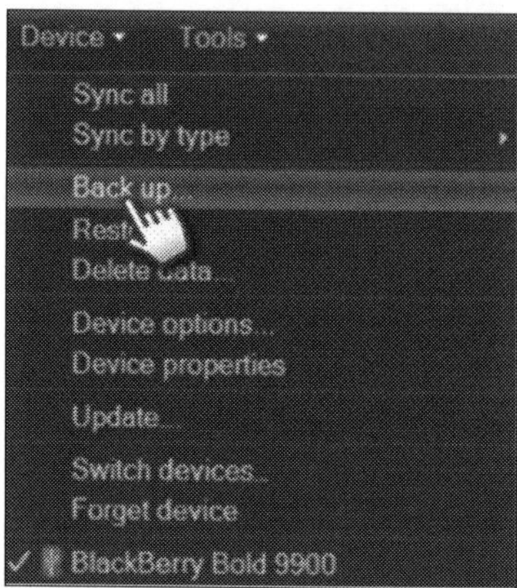

4. Select **Backup type** as **Full (all device data and settings)** to perform a full backup, as shown in the following screenshot:

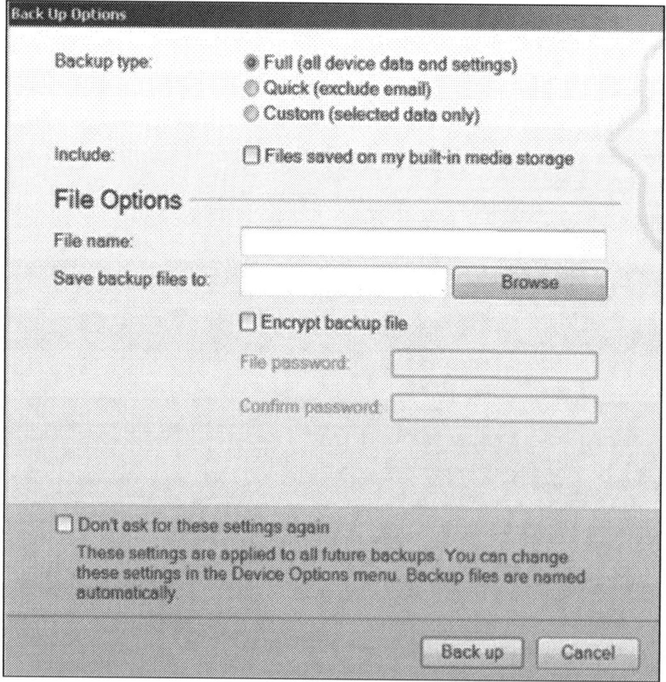

Full backup option in BlackBerry

5. As shown in the preceding screenshot, the **File name** and location to save the backup file must be selected. You are recommended to name the file accordingly to reflect the naming convention implemented by your organization or to simply use the device name and serial number. This will ensure that the backup file can easily be associated back to the original device. Once this is complete, click on **Back up**.

BlackBerry analysis

BlackBerry devices are still used by employees of major corporations due to the great security features. **eDiscovery** cases often require the examiner to be well versed in extracting and analyzing data from computers, servers, and smartphones such as BlackBerry devices. Commercial tools are available for the analysis of BlackBerry devices. The method of acquisition will determine the amount of analysis possible by the examiner. For example, a physical acquisition may have been obtained, but the forensic tool does not automatically parse the data in the image file. This requires the examiner to manually carve and reconstruct the data. BlackBerry devices are one of the most complicated smartphones to understand and consistently reconstruct by manual examination. The previous section provided some steps to successfully extracting data from BlackBerry devices. The acquisition steps should be followed to ensure that data is not missed. Multiple acquisitions may be required in order to extract and recover the user data from a BlackBerry device. The methodologies and forensic tools required to analyze data from BlackBerry backup files and forensic images differ, and they are defined in the following sections.

BlackBerry backup analysis

BlackBerry backup files can be found natively on hard drives or other external media during a forensic investigation or may exist as the *forensic image* created by the examiner in order to complete their forensic investigation. Sometimes, the backup file contains more *usable* data than a physical image. Again, it all depends on the device model, the method of acquisition and the forensic tool used for analysis. As previously mentioned, BlackBerry backup files exist as IPD and BBB files and are created by the BDM or the BlackBerry Link software. When created by a user, the BlackBerry backup files are commonly stored in the `My Documents` folder on a Windows platform. The backup file contains various databases (tables) present on the BlackBerry device. It is named by default in the format `Backup (yyyy-mm-dd).ipd`.

Best practices suggest searching for IPD and BBB files across digital media suspected of containing BlackBerry backup files since the user can modify the filename of the backup. If the BlackBerry backup file was recovered from a hard drive or other digital media, the following two formats may exist:

- `Loaderbackup (yyyy-mm-dd).ipd`
- `AutoBackup ((yyyy-mm-dd).ipd`

The `Loaderbackup` file is created automatically when the device OS is being updated. This ensures that required data is readily available should the device crash during the upgrade. The `Autobackup` file is created when the user elects to have the device set to back up on a regular or scheduled basis or when the device is synced with a PC.

A full backup of a BlackBerry device should contain details such as address book, e-mail, SMS, call logs, and more. However, the backup file may not contain all the application data because the third-party applications may not always provide access to their data. A backup file contains the following information:

- **File header**: The header contains information about the RIM signature, database version, number of databases in the current file, and so on, as shown in the following table:

Name	Length (in bytes)	Offset
RIM signature	37	`0x0`
Line break	1	`0x25`
Database version	1	`0x26`
Number of databases	2	`0x27~0x28`
Database separator	1	`0x29`

- **Database name blocks**: These are present after the header information. In each block, the name length and name are stored.
- **Database records**: These are present after database name blocks and contain real data. They contain information about database ID, record length, database version, database record handle, database unique ID, and so on.
- **Database record fields**: These contain record data length, record type, and record data.

Once you have access to the BlackBerry backup file, use any of the available tools discussed in the *Forensic tools for BlackBerry analysis* section to read the information present in the file.

BlackBerry forensic image analysis

The method of obtaining the forensic image of a BlackBerry device, whether logical, physical, or file system, may limit the tools available to analyze the data. For example, a raw image created using JTAG or chip-off should be ingestible and parsed by any forensic tool that provides physical analysis support for that model of BlackBerry, as long as the device was unlocked or the passcode is known. It is best to use more than one tool during your forensic analysis to verify the results of the forensic image.

BlackBerry file systems are difficult to reconstruct due to the proprietary format developed by RIM. Unlike other smartphone devices, BlackBerry file systems vary greatly per model. Commercial tools will attempt to reconstruct the file systems, but the support is low and may not be accurate. It is best to validate your findings using logical, file system, or backup file acquisition and analysis to ensure your findings are correct.

Once an examiner gains experience analyzing BlackBerry devices, the files of interest become more apparent regardless of the image format. A physical dump and backup file may actually contain the same amount of data readily available to the examiner. The tool of choice to examine the data will determine the amount of access you have to that file. As explained in previous chapters, deleted data can reside in database files just as Android and iOS, BlackBerry databases/tables may contain deleted data. If your forensic tool does not provide access to the native file for export or for examination in Hex, you will miss this deleted data.

The following screenshot shows the file system representation of a BlackBerry backup file in Cellebrite Physical Analyzer. Notice that the **Address Book** is being examined in *raw hex*. This method of analysis is preferred to validate your logical results or the data provided in the tool report.

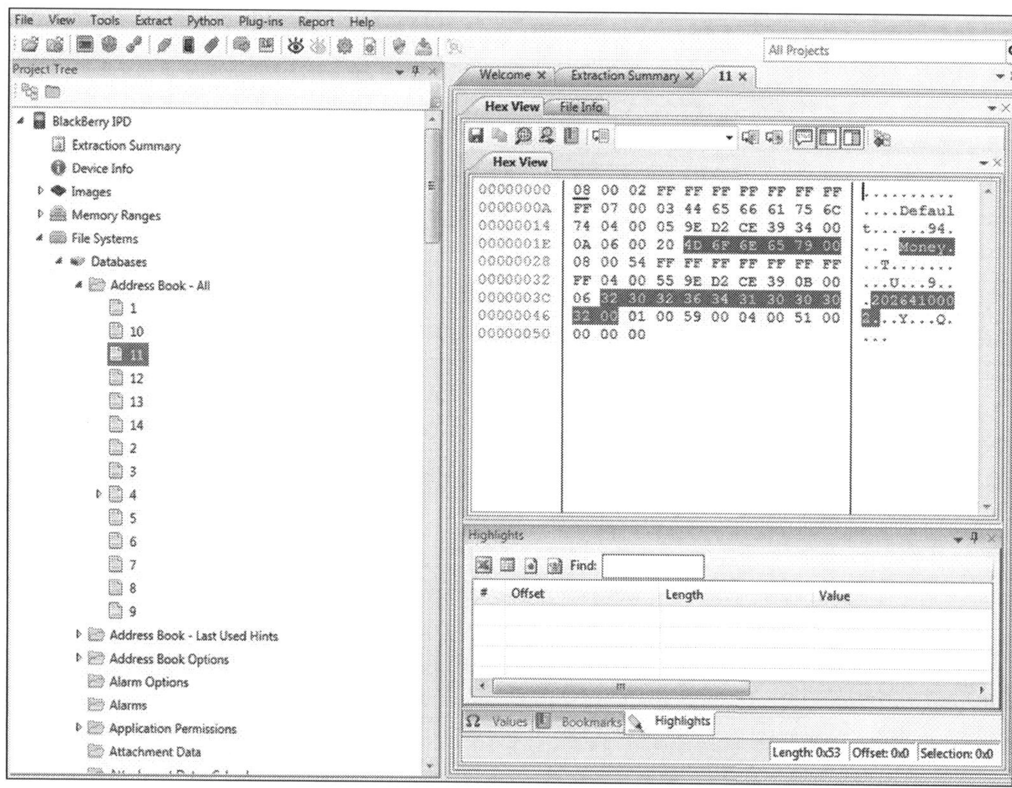

Cellebrite Physical Analyzer – Address Book examination

BlackBerry data, stored in `databases/tables`, is often proprietary, which causes difficulties when attempting to interpret data using the tool and manually by the examiner. When compared to other smartphone devices, there doesn't appear to be a clear standard for data on BlackBerry devices. For example, status flags associated with the e-mail app have been found to be inconsistent among different devices. Commonly, a status flag is consistent within a table for a specific model. This has been found to be untrue for BlackBerry. For examiners, this makes validation of your tool difficult. BlackBerry timestamps are commonly in a simple date format, which is compatible with Java and is supported to parse by most forensic tools.

There are a variety of BlackBerry timestamp types that are defined in detail at `http://www.swiftforensics.com/2012/03/blackberry-date-formats.html`. When examining SMS messages, the examiner should use more than one tool to ensure the data is parsed properly. Currently, there is no standard for how SMS messages are stored on BlackBerry devices. The SMS messages may be encrypted, compressed, or exist as a proprietary 7-bit format. Several factors weigh on the format to store the SMS message content, including device security settings, device model, administrator settings, and more.

Unlike other smartphones, third-party application data cannot be stored internally on the BlackBerry device memory if the application uses SQLite database storage, which applications commonly do. All third-party application data will reside on the SD card (or eMMC) associated to the BlackBerry device in an application folder. More information on using SQLite on BlackBerry devices can be found at `http://blog.softartisans.com/2011/03/29/using-sqlite-in-blackberry-applications/`. These folders and database files must be examined for relevance to the investigation, as defined in previous chapters. Due to the unknown nature of RIM and the proprietary methods to store user data, it is recommended that the examiner examine any database/table recovered from the BlackBerry device that may be of interest to the investigation. Manual examination is time-consuming, but it will ensure that data is not overlooked.

Encrypted BlackBerry backup files

During your forensic examinations, it is likely that an encrypted BlackBerry backup file will require analysis. Elcomsoft developed the Phone Password Breaker, which allows the examiner to use various brute force and dictionary attacks to crack encrypted backup files.

The following is the step-by-step procedure to crack an encrypted BlackBerry backup file using **Elcomsoft Phone Password Breaker**:

1. On the forensic workstation, install Elcomsoft Phone Password Breaker. The full and demo version can be found at `http://www.elcomsoft.com/eppb.html`.

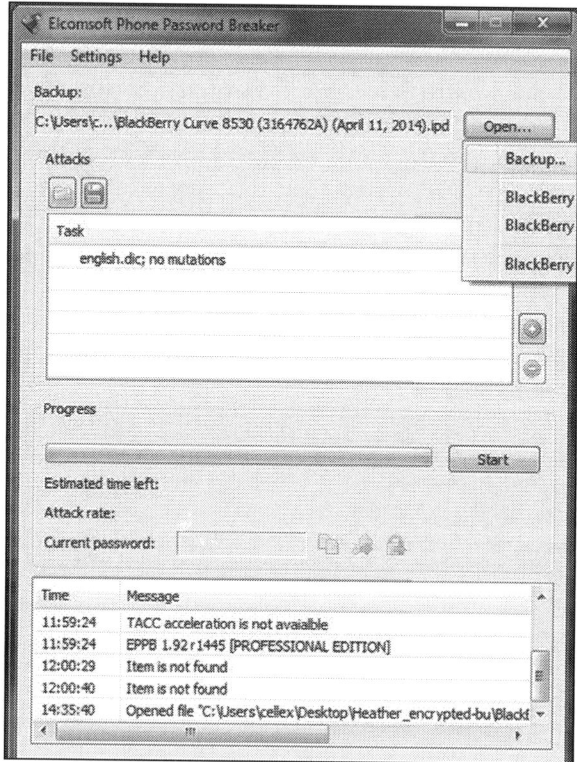

Elcomsoft Phone Password Breaker

2. Navigate to the backup file.

3. Select the attack method. Several options are available and dictionaries can be added to increase the success rate of the attack, as shown in the following screenshot:

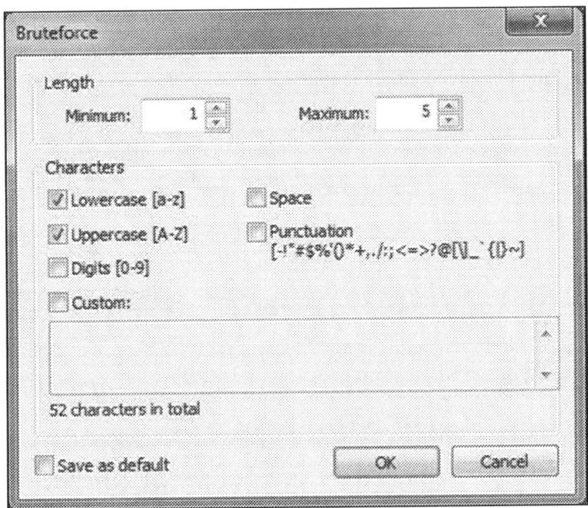

The Elcomsoft Phone Password Breaker attack options

4. If cracked, the password will be displayed and can be used to access the encrypted backup file with the use of a forensic tool. It is important to use a forensic tool that will prompt you for the password. Some will simply fail or finish with errors and provide no access to the encrypted data, as shown in the following screenshot:

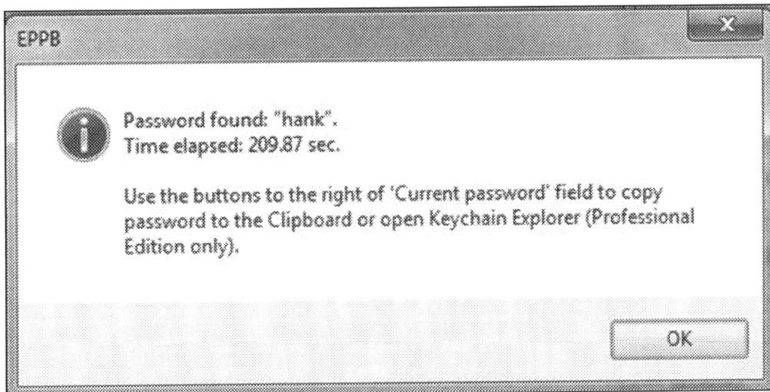

Elcomsoft Phone Password Breaker

Forensic tools for BlackBerry analysis

Several forensic tools are available to parse data from BlackBerry backup files and forensic images of BlackBerry devices. The best tools should provide access to the raw database files to ensure that data not supported by the forensic tool could be manually parsed by the examiner and to avoid deleted data not being recovered. Knowing where to find the data on devices takes practice and the examiner should be trained on examining data from BlackBerry devices.

Some forensic tools available include Cellebrite Physical Analyzer, Oxygen Forensics Suite, Microsystemation XRY, AccessData MPE+, and several others. Some tools are specifically designed to analyze BlackBerry backup files. Common tools that provide support for backup files include Oxygen Forensics IPD Viewer, Elcomsoft BlackBerry Backup Explorer, and BlackBerry Backup Extractor. Bulk Extractor, created by Dr. Simson Garfinkle, is a free tool that can parse data from raw BlackBerry image files (physical dumps) even if the password is unknown.

Bulk Extractor scans the image file and pulls useful information (calls, URLs, e-mail addresses, and more) without parsing the file system and provides the results to the examiner. Bulk Extractor can be downloaded from `http://digitalcorpora.org/downloads/bulk_extractor/`. An example of a Bulk Extractor output for telephone numbers is shown in the following screenshot:

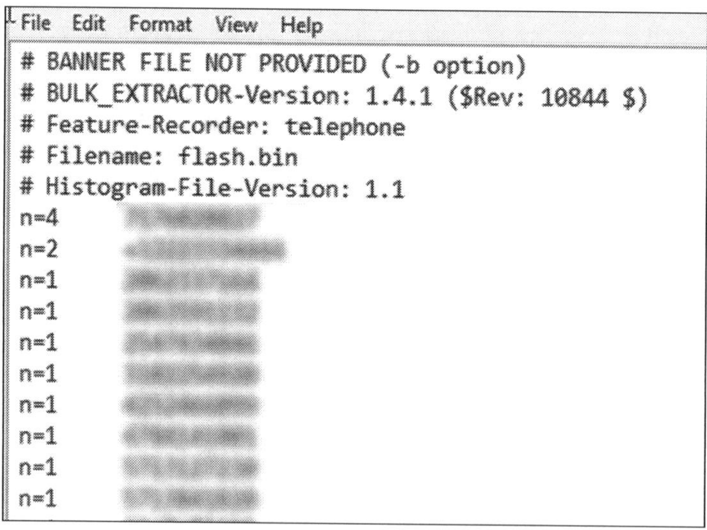

Telephone numbers parsed by Bulk Extractor

The following is a step-by-step procedure to view the information present in an IPD file using BlackBerry Backup Extractor. This tool provides access to the native files for further examination. A tool such as BlackBerry Backup Extractor may be helpful when your commercial forensic tool does not provide access to the actual files recovered from the BlackBerry backup file.

1. Download and install BlackBerry Backup Extractor on the forensic workstation (`http://www.blackberryconverter.com/`).
2. Click on the **Open backup...** button to load the IPD backup file into the software as shown in the following screenshot:

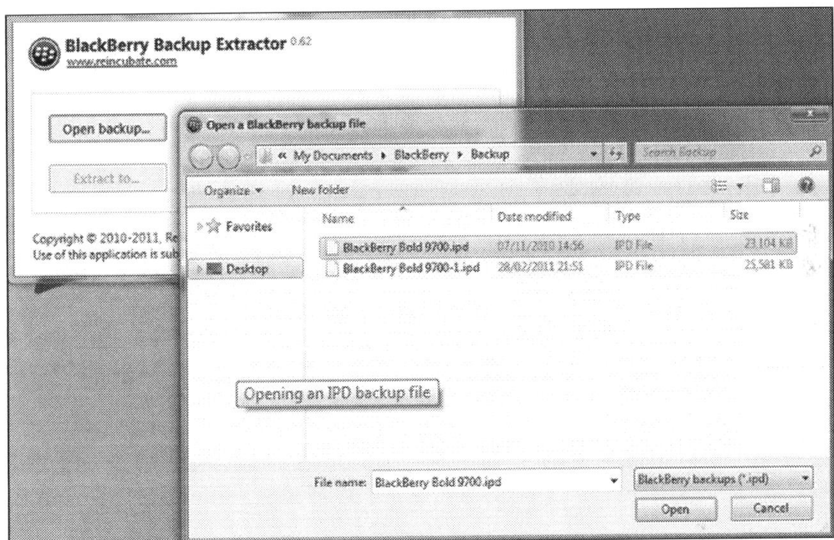

BlackBerry Backup Extractor

3. Select the folder where the data will be saved and extracted. When the process begins, the tools display information about the number of databases currently being extracted.

4. Once the extraction is complete, you will find information about sent e-mails, received e-mails, contacts, SMS, calendar appointments, and more, as shown in the following screenshot:

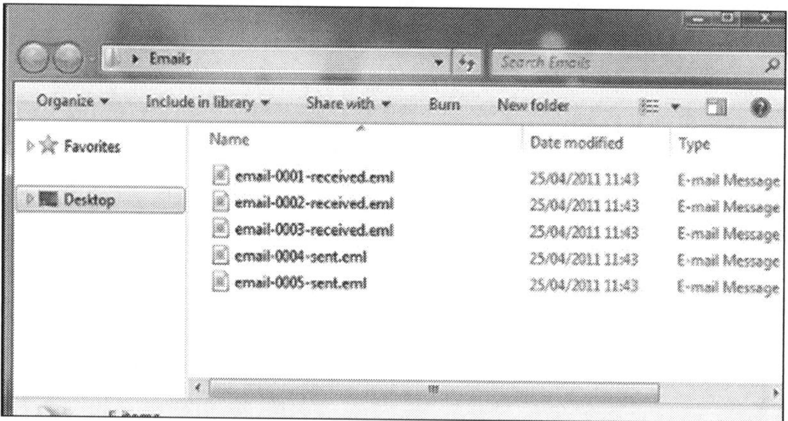

E-mail extracted from backup

The contacts, call logs, and other data extracted by the tool can be navigated to and examined for relevance as shown in the following screenshot. Again, BlackBerry Backup Extractor does not provide an analytical platform to view all of the extracted data in a normalized manner; therefore, manual review of the results is required.

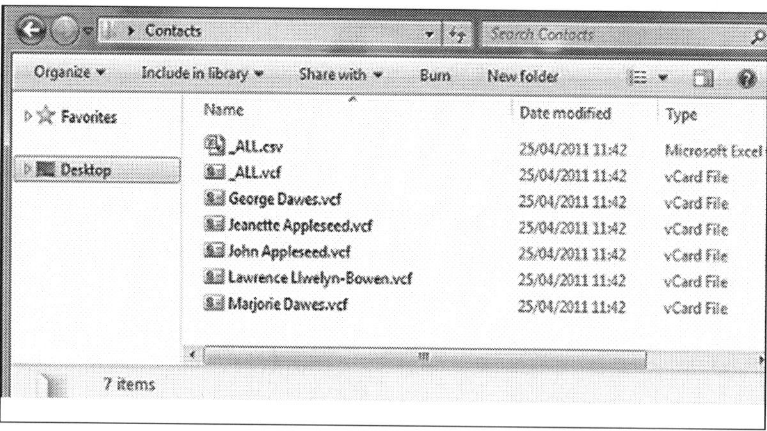

Contacts extracted from backup

Other information that can be crucial during investigations, such as browser URLs, browser data cache, and so on, are also extracted as shown in the following screenshot:

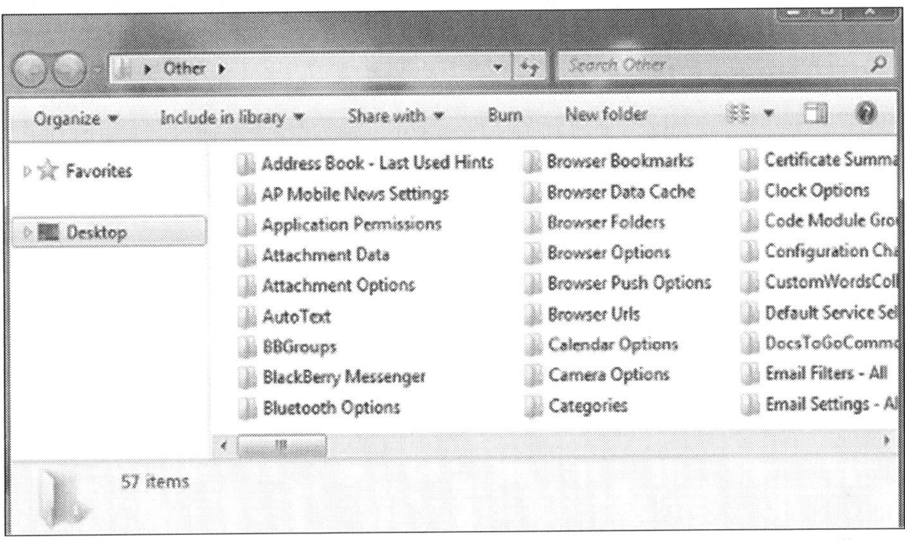

Other useful data extracted from the backup

The BlackBerry backup file contains a 2-byte hexadecimal value that, when converted to decimal, reveals the number of database files contained within that backup file. The two bytes of interest are the third and fourth bytes following the file header of the IPD backup file. As shown in the following screenshot, the IPD file is being examined in a Hex viewer to determine the number of database files contained within the IPD file. The third and fourth bytes (`00 6D`) are going to be converted for database verification purposes. In the following screenshot, Hex `6D` is converted to decimal, which is `109`. Therefore, there are 109 databases contained within this IPD file. It is important for the forensic tool to display 109 databases/tables for the examiner to analyze.

The Hex view of IPD file

Some forensic tools will convert this number for you, which is true with Oxygen Forensics IPD Viewer, as shown in the next screenshot. Oxygen Forensics Suite is one of the most powerful commercial forensic tools to parse data from BlackBerry backup files. This suite of tools offers both a backup file parser as well as an IPD Viewer. Some forensic tools omit empty databases, provide partial support for backup files, or require the examiner to manually convert and verify the number of tables. To verify the number of databases/tables in a BlackBerry backup file, Elcomsoft IPD Viewer can be used by performing the following steps:

1. Install Oxygen Forensics Suite (license required) on your forensic workstation.
2. Select BlackBerry IPD Viewer and navigate to the backup file. Have a look at the following screenshot:

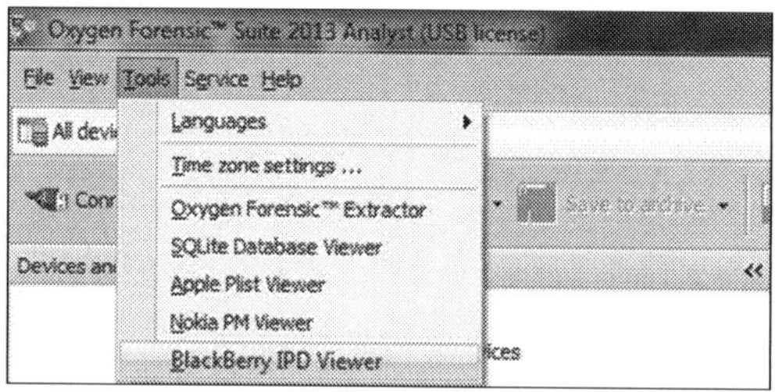

Oxygen Forensics Suite BlackBerry IPD Viewer

3. If encrypted, enter the password. If you do not know the password, the data cannot be decrypted for examination. Keep in mind that you will need this password every time you open the image file for examination, as shown in the following screenshot:

Oxygen Forensics Suite BlackBerry IPD Viewer—the encrypted file

4. The decrypted data will be provided for examination. Note that the number of databases contained within this backup file was `107` as shown in the following screenshot:

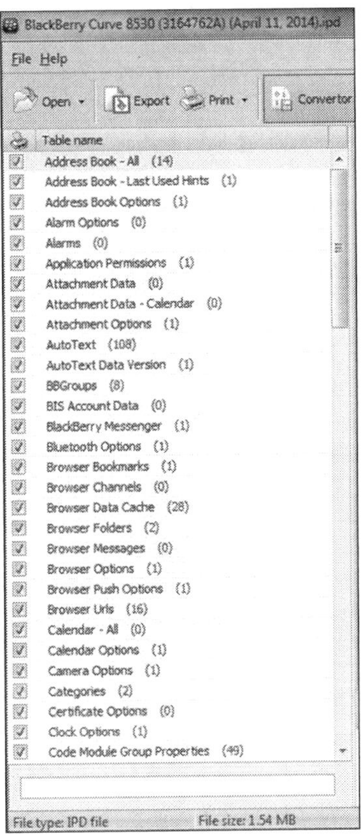

The Oxygen Forensics Suite BlackBerry IPD Viewer results

Summary

Forensic support for BlackBerry devices is limited when compared to other smartphone devices. Open source tools supporting BlackBerry physical acquisition are not currently available, and bypassing a locked device is complicated and often renders the data encrypted and unusable. Unlike iOS and Android devices, our most successful data extractions of BlackBerry devices usually come from the file system image or backup file. Information such as e-mail, SMS, contacts, and more can be extracted from BlackBerry backup files. Sometimes, the most useful information is the data extracted from a backup file, which provides access to the most data for analysis.

Index

Symbols

A

B

C

D

E

F

G

J

K

L

M

N

O

P

Q

R

S

T

U

V

W

Y

Z

Thank you for buying
Practical Mobile Forensics

About Packt Publishing

Packt, pronounced 'packed', published its first book "*Mastering phpMyAdmin for Effective MySQL Management*" in April 2004 and subsequently continued to specialize in publishing highly focused books on specific technologies and solutions.

Our books and publications share the experiences of your fellow IT professionals in adapting and customizing today's systems, applications, and frameworks. Our solution based books give you the knowledge and power to customize the software and technologies you're using to get the job done. Packt books are more specific and less general than the IT books you have seen in the past. Our unique business model allows us to bring you more focused information, giving you more of what you need to know, and less of what you don't.

Packt is a modern, yet unique publishing company, which focuses on producing quality, cutting-edge books for communities of developers, administrators, and newbies alike. For more information, please visit our website: `www.packtpub.com`.

About Packt Open Source

In 2010, Packt launched two new brands, Packt Open Source and Packt Enterprise, in order to continue its focus on specialization. This book is part of the Packt Open Source brand, home to books published on software built around Open Source licenses, and offering information to anybody from advanced developers to budding web designers. The Open Source brand also runs Packt's Open Source Royalty Scheme, by which Packt gives a royalty to each Open Source project about whose software a book is sold.

Writing for Packt

We welcome all inquiries from people who are interested in authoring. Book proposals should be sent to `author@packtpub.com`. If your book idea is still at an early stage and you would like to discuss it first before writing a formal book proposal, contact us; one of our commissioning editors will get in touch with you.

We're not just looking for published authors; if you have strong technical skills but no writing experience, our experienced editors can help you develop a writing career, or simply get some additional reward for your expertise.

Metasploit Penetration Testing Cookbook

Second Edition

ISBN: 978-1-78216-678-8 Paperback: 320 pages

Over 80 recipes to master the most widely used penetration testing framework

1. Special focus on the latest operating systems, exploits, and penetration testing techniques for wireless, VOIP, and cloud.
2. This book covers a detailed analysis of third-party tools based on the Metasploit framework to enhance the penetration testing experience.

Web Penetration Testing with Kali Linux

ISBN: 978-1-78216-316-9 Paperback: 342 pages

A practical guide to implementing penetration testing strategies on websites, web applications, and standard web protocols with Kali Linux

1. Learn key reconnaissance concepts needed as a penetration tester.
2. Attack and exploit key features, authentication, and sessions on web applications.
3. Learn how to protect systems, write reports, and sell web penetration testing services.

Please check **www.PacktPub.com** for information on our titles

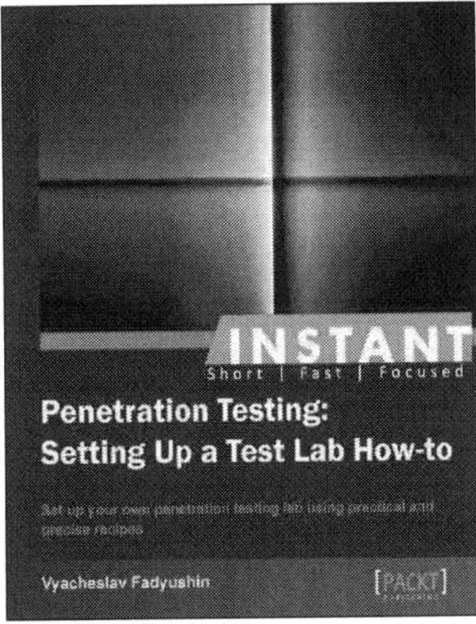

Instant Penetration Testing: Setting Up a Test Lab How-to

ISBN: 978-1-84969-412-4 Paperback: 88 pages

Set up your own penetration testing lab using practical and precise recipes

1. Learn something new in an Instant! A short, fast, focused guide delivering immediate results.
2. A concise and clear explanation of penetration testing, and how you can benefit from it.
3. Understand the architectural underpinnings of your penetration test lab.

Kali Linux Cookbook

ISBN: 978-1-78328-959-2 Paperback: 260 pages

Over 70 recipes to help you master Kali Linux for effective penetration security testing

1. Recipes designed to educate you extensively on the penetration testing principles and Kali Linux tools.
2. Learning to use Kali Linux tools, such as Metasploit, Wire Shark, and many more through in-depth and structured instructions.
3. Teaching you in an easy-to-follow style, full of examples, illustrations, and tips that will suit experts and novices alike.

Please check **www.PacktPub.com** for information on our titles

28306047R00184

Made in the USA
San Bernardino, CA
27 December 2015